I Stewart
University of Warwick

Lie algebras generated by finite-dimensional ideals

Pitman Publishing
LONDON · SAN FRANCISCO · MELBOURNE

AMS Subject Classifications: 17B65, 17B05, 20G99, 20E25

PITMAN PUBLISHING
Pitman House, 39 Parker Street, London WC2B 5PB, UK

PITMAN PUBLISHING CORPORATION
6 Davis Drive, Belmont, California 94002, USA

PITMAN PUBLISHING PTY LTD
Pitman House, 158 Bouverie Street, Carlton, Victoria 3053, Australia

PITMAN PUBLISHING
COPP CLARK PUBLISHING
517 Wellington Street West, Toronto M5V 1G1, Canada

SIR ISAAC PITMAN AND SONS LTD
Banda Street, PO Box 46038, Nairobi, Kenya

PITMAN PUBLISHING CO SA (PTY) LTD
Craighall Mews, Jan Smuts Avenue, Craighall Park,
Johannesburg 2001, South Africa

ISBN 273 00142 6

Reproduced and printed by photolithography and bound in Great Britain at The Pitman Press, Bath

1707 : 97

Contents

Introduction

These notes are intended as a contribution to the structure theory of infinite-dimensional Lie algebras from two points of view. First, as a generalization of the classical theory of Killing, Cartan, Mal'cev, Levi, and others, on the structure of finite-dimensional Lie algebras over algebraically closed fields of characteristic zero. Second, as a step towards a 'formation theory' of infinite-dimensional Lie algebras, analogous to existing theories for various classes of finite or infinite groups. In a nutshell, the former suggests what theorems to prove and the latter suggests how to prove them. The following brief history of formation theory may help to set the notes in their proper context.

With hindsight, the earliest result in formation theory is Sylow's theorem on the conjugacy of maximal p-subgroups of finite groups, published in 1872. The next significant work is that of Philip Hall [22] in 1928, in which he proved that if π is any set of primes, then every finite soluble group has a unique conjugacy class of maximal π-subgroups. Papers of Hall published in 1937 [23,24] and 1956 [25] extended these ideas. However, the arithmetical nature of these theorems obscures a more structural approach leading to a wide range of related phenomena. In 1961 Roger Carter [12] proved that every finite soluble group has a unique conjugacy class of nilpotent self-normalizing subgroups. The analogy with Cartan subalgebras of Lie algebras is clear: not surprisingly these subgroups were immediately christened <u>Carter subgroups</u>. The inductive nature of Carter's proof, and of Hall's, led to the discovery by Gaschütz [19] in 1963 that both ideas are special cases of a general concept. He defined a <u>formation</u> of finite soluble groups to be a class of finite soluble groups $\mathfrak{F}$, closed under isomorphism, with the further properties:

If $G \in \mathfrak{F}$ and $H \triangleleft G$ then $G/H \in \mathfrak{F}$, (1)

If $H, K \triangleleft G$, and both G/H and $G/K \in \mathfrak{F}$, then $G/(H \cap K) \in \mathfrak{F}$. (2)

Further, $\mathfrak{F}$ is <u>saturated</u> if

$G/\Phi(G) \in \mathfrak{F}$ implies $G \in \mathfrak{F}$, (3)

where $\Phi(G)$ is the Frattini subgroup. The main result of Gaschütz was that if $\mathfrak{F}$ is any saturated formation and G any finite soluble group, then G has a unique conjugacy class of $\mathfrak{F}$-<u>projectors</u>. These latter are defined to be subgroups P of G with the properties:

$P \in \mathfrak{F}$, (4)

Whenever H, K are subgroups of G such that $P \leq H \triangleright K$ and $H/K \in \mathfrak{F}$, then $H = KP$. (5)

(We may paraphrase (5) as: P <u>covers</u> every $\mathfrak{F}$-section above P.)

By taking $\mathfrak{F}$ to be the class $\mathfrak{F}_p$ of finite p-groups one obtains as $\mathfrak{F}_p$-projectors the Sylow p-subgroups of G. If $\mathfrak{F} = \mathfrak{F}_\pi$, the class of finite π-groups for a set of primes π, then the $\mathfrak{F}_\pi$-projectors are the Hall π-subgroups. For the class $\mathfrak{N}$ of finite nilpotent groups, the $\mathfrak{N}$-projectors are the Carter subgroups.

Formation theory of finite soluble groups has developed into a large area of research, at the hands of many mathematicians. Two offshoots of this work are relevant in the present context. Barnes and Gastineau-Hills [6] showed how to obtain an analogous formation theory of finite-dimensional soluble Lie algebras. If $\mathfrak{N}$ denotes the class of nilpotent Lie algebras the $\mathfrak{N}$-projectors are the original Cartan subalgebras. This subject also developed rapidly, with contributions from Barnes [3], Barnes and Newell [7], Stitzinger [66,67]. Furthermore, in certain cases the solubility assumption may be dispensed with. Let $\mathfrak{N}$, $\mathfrak{B}$, and $\mathfrak{S}$ be the classes of nilpotent, soluble, and semisimple Lie algebras of finite dimension over an algebraically closed field of characteristic zero. If L is a finite-dimensional Lie algebra then the $\mathfrak{N}$-projectors of L are its Cartan subalgebras; the $\mathfrak{B}$-projectors its Borel subalgebras (maximal soluble subalgebras); and the $\mathfrak{S}$-projectors its Levi subalgebras (maximal semisimple subalgebras, classically called Levi factors since they complement the soluble radical). In each case there is a classical conjugacy theorem (under a certain group of automorphisms of L).

The other offshoot involves the generalization of formation theory of finite soluble groups to various classes of infinite groups. The Sylow and Hall theory had already been generalized in this way, with especial success for <u>periodic FC-groups</u> (which may be defined as those groups generated by a set of finite normal subgroups) by Baer [2] and Gol'berg [20]. Here conjugacy is replaced by local conjugacy: two subgroups are locally conjugate if one can be mapped to the other by an automorphism whose restriction to any finite subset is equal to the restriction of some inner automorphism. Such automorphisms are constructed using projective limits, and the theorem that a projective limit of non-empty finite sets is non-empty (known to many group theorists as a theorem on 'projection sets' of Kuroš [45] p.167). Stonehewer [68] obtained a theory of Carter subgroups in certain locally finite groups, and in [69] extended his results to periodic FC-groups. In [71] Tomkinson developed a formation theory for locally soluble periodic FC-groups. It is this paper which provided much of the motivation for the present work. Wehrfritz [76] discussed Sylow subgroups of linear groups, and in [77] dealt with Carter subgroups and basis normalizers of linear groups. A certain amount of unification was achieved by Hartley, Gardiner, and Tomkinson [37] who invented a class $\mathfrak{U}$ of locally finite locally soluble groups with 'good' Sylow theory, and inferred 'good' formation theory. The class $\mathfrak{U}$ includes the groups studied by Stonehewer [68] and Wehrfritz, but does not include FC-groups. Its properties have been extensively investigated by Hartley [28,29,30,31,32,33,34,35] and by Graddon and Hartley [21]. Klimowicz [42,43,44] developed a general axiomatic setting which included $\mathfrak{U}$-groups and FC-groups, showing (roughly speaking) that if to each group in a suitable class there is associated, in a 'functorial' way, a group of automorphisms, and if this group is transitive on the Sylow structure, then it is transitive on the formation structure.

Recently Newell [49, 50] has discovered a kind of 'formation theory modulo things of finite index', generalizing the formation concept to that of a <u>format</u>. This theory is particularly appropriate to polycyclic groups. (It also seems very closely related in spirit to Chevalley's abstract definition of the Cartan subgroups of an algebraic group, to be found in Chevalley [14] p.279 or Borel [9] pp.271,290. Perhaps this is a starting point for a formation theory of algebraic groups or Lie groups.)

There is also a 'dual' theory of <u>Fitting classes</u> and <u>injectors</u>, due to Fischer [17] and developed by Fischer, Gaschütz and Hartley [18], Hartley [36], and recent papers of Dark, Doerk, Hawkes, Lockett, and others. Tomkinson [72] discusses Fitting classes of FC-groups. For Lie algebras this dual theory seems less interesting, and nothing has been published. What little I know about the matter has been collected in the appendix at the end of these notes.

The interactions involved in this development may be represented schematically by the diagram:

FINITE GROUPS ⟶ INFINITE GROUPS

↑ ↓

FINITE-DIMENSIONAL
LIE ALGEBRAS

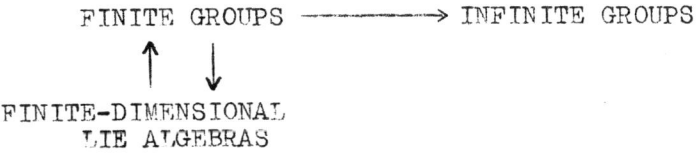

An obvious thing to attempt, especially for someone with my interests, is to construct the 'pushout':

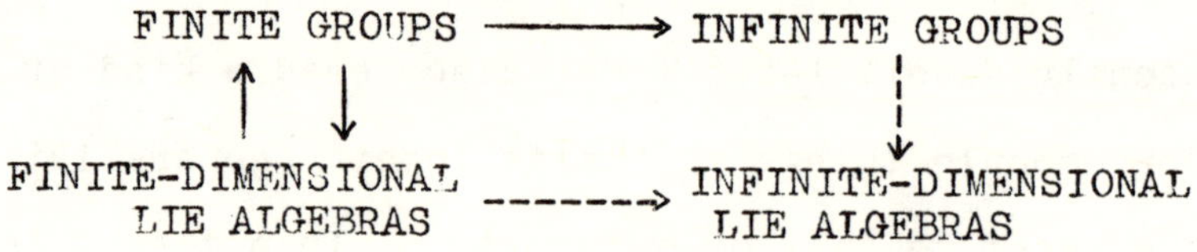

For several reasons the most attractive class of infinite-dimensional Lie algebras to consider is that analogous to the class of periodic FC-groups: namely, the class of Lie algebras generated by a set of finite-dimensional ideals. (It is in any case unclear how one might usefully generalize linear groups or $\mathfrak{U}$-groups.) Algebras in this class are very special sorts of locally finite Lie algebras, so there is some hope of gaining insight into locally finite Lie algebras, a fascinating subject about which little is known. The experience in group theory is that although FC-groups are far from typical, in that they are too well behaved, nevertheless they are useful as a platform from which to attack locally finite groups (see Kegel and Wehrfritz [41]).

Technical reasons make it necessary (for me!) to work over a field which is algebraically closed and of characteristic zero. This is a natural restriction for conjugacy theorems, in that it is required even in finite dimensions, but it is less natural for existence theorems. So much is known about finite-dimensional Lie algebras over such fields that we may dispense with any solubility assumptions (unlike the situation in group theory), which at least is some compensation for the restrictions on the field. However, what we get is not a fully-fledged formation theory. There is a reason for this. In finite dimensions, and over fields of the above type, it has been shown by Barnes and Gastineau-Hills [6] that the only saturated formations of soluble Lie algebras are the zero-dimensional, nilpotent, or soluble Lie algebras. Even if the solubility assumption on the algebra is removed, this means that the projectors for saturated formations of soluble algebras are the Cartan subalgebras (nilpotent projectors) and Borel subalgebras (soluble projectors). Another insoluble formation for which projectors exist is the class of semisimple algebras, where the projectors are the Levi factors (henceforth called Levi subalgebras for uniformity). Thus what we emerge with is a generalization of the classical structure theory of finite-dimensional Lie algebras, with emphasis on these three special formations, rather than a general theory of arbitrary formations. In fact the generalization is quite extensive and the special nature of the results is no great disappointment: on the contrary, the wealth of detail available in the case of Cartan, Borel, and Levi subalgebras more than compensates for a loss of generality which is, in any case, somewhat illusory. Formation theory remains prominent, however, in the _proofs_ of the main theorems.

Most of the results in this set of notes were worked out at the University of Tübingen in 1974, with the aid of a grant from the Alexander von Humboldt - Stiftung. The results on radicals and Levi subalgebras are older, and are special cases of the results of Amayo and Stewart [1] chapter 13 pp.256-273. Large sections have been published, under various titles, as [54,55,56,57,58,59]. However, several theorems have not hitherto been published; and a comprehensive account may prove useful. (Incidentally, the entire set of notes may be construed as an answer to part of question 33, p.396, of [1].) One thing which is clear is that the theory can be taken much further than is done here, and similar results obtained in other areas. In particular, associative algebras can be subjected to the same treatment [60].

Many mathematicians have contributed to the development of this theory. Brian Hartley and Stewart Stonehewer have explained various points in group theory and indicated lines of attack on the Lie algebra analogues. Roger Richardson and David Winter provided useful information on automorphism groups of Lie algebras and their structure as algebraic groups. David Mumford and Pierre Deligne were kind enough to answer my queries about projective limits in algebraic geometry. It is a pleasure to record my thanks for these invaluable contributions.

The broad line of development is as follows. The first two chapters give resumés of relevant results on finite-dimensional Lie algebras and affine algebraic groups (treated in as simple-minded a fashion as possible). Next we introduce the class of <u>locally finite</u> Lie algebras and certain distinguished subclasses, of which the crucial one is the class of <u>ideally finite</u> Lie algebras (analogous to the class of FC-groups).

Next we develop the elementary parts of the theory: locally nilpotent and locally soluble radicals, the existence of Levi subalgebras, the Frattini subalgebra. (The results on the Frattini subalgebra have not been published before, and are consequently developed in a more general context which may be of interest.) The techniques here are Zorn's lemma, 'local nonsense' (consider a local system of subalgebras, ideals, etc. with suitable properties), and the dual 'residual nonsense' (argue on a system of quotient algebras).

The way to conjugacy theorems is paved by a projective limit theorem for certain algebraic varieties derived from algebraic groups. From this we derive conjugacy theorems for Levi, Borel, and Cartan subalgebras, and an existence theorem for Cartan subalgebras. (The existence of Borel subalgebras is trivial). These subalgebras are then investigated with a view to generalizing standard properties from finite dimensions.

Finally the Fitting and Chevalley-Jordan decompositions are invoked (weight spaces and nilpotent-semisimple splitting) to yield useful characterizations of Cartan subalgebras, an alternative proof of their existence, results on the toral structure, and a generalization to ideally finite Lie algebras of Mal'cev's results [46] on 'splittable' Lie algebras including a classification of maximal locally nilpotent subalgebras.

1 Résumé: Lie algebras

Throughout these notes, k will always denote an algebraically closed field of characteristic zero. Although many of the results we shall prove are true under less restrictive hypotheses we shall not attempt maximum generality.

The object of this chapter is to state explicitly those facts about finite-dimensional Lie algebras over k which are needed in the sequel, either as motivation or as raw material for proofs. In the process we shall settle details of notation. A certain amount of familiarity with this material is assumed. We also discuss some elementary results on Lie algebras of infinite dimension. For more details the reader should refer to Humphreys [38], Jacobson [39], Kaplansky [40], Samelson [51], Winter [78], or (for infinite dimensions) Amayo and Stewart [1].

Let L be a Lie algebra over k. If H is a subalgebra of L we write $H \leq L$, and if H is an ideal we write $H \triangleleft L$. We use square brackets [,] to denote Lie multiplication in L. If A, B are subspaces of L we write $[A,B]$ for the subspace spanned by all $[a,b]$ for $a \in A$, $b \in B$. If X is a subset of L then $\langle X \rangle$ (resp. $\langle X^L \rangle$) is the subalgebra (ideal) generated by X.

The <u>derived</u>, <u>lower central</u>, and <u>upper central series</u> of L are defined recursively by

$$L^{(0)} = L, \quad L^{(n+1)} = [L^{(n)}, L^{(n)}]$$

$$L^{1} = L, \quad L^{n+1} = [L^{n}, L]$$

$$\zeta_1(L) = \{x \in L : [L,x] = 0\}, \quad \zeta_{n+1}(L)/\zeta_n(L) = \zeta_1(L/\zeta_n(L)).$$

If L has infinite dimension these series may be continued transfinitely by taking n to be any ordinal, and defining for limit ordinals λ

$$L^{(\lambda)} = \bigcap_{n<\lambda} L^{(n)}, \quad L^{\lambda} = \bigcap_{\lambda<n} L^{n}, \quad \zeta_{\lambda}(L) = \bigcup_{n<\lambda} \zeta_n(L).$$

We say L is <u>soluble</u> if (for finite n) $L^{(n)} = 0$ for some n, and L is <u>nilpotent</u> if (for finite n) $L^n = 0$. Note that each of $L^{(n)}$, L^n, $\zeta_n(L)$ is an ideal of L.

If X is a subset of L we define the <u>centralizer</u>

$$C_L(X) = \{x \in L: [X,x] = 0\},$$

and the <u>idealizer</u>

$$I_L(X) = \{x \in L: [X,x] \subseteq X\}.$$

The latter is a subalgebra if X is a subspace; the former is always a subalgebra. Likewise, if M is an L-module, we write

$$C_L(M) = \{x \in L: Mx = 0\}.$$

This is an ideal, and is the kernel of the representation associated with M. If $C_L(M) = 0$ we say that M is <u>faithful</u>.

A <u>derivation</u> of a Lie algebra L is a linear map $\delta: L \to L$ such that

$$[x,y]\delta = [x\delta, y] + [x, y\delta] \qquad (x,y \in L).$$

The set of all derivations of L forms a Lie algebra under commutation $\delta_1\delta_2 - \delta_2\delta_1 = [\delta_1, \delta_2]$, which we write as Der(L). There is a homomorphism $*: L \to \mathrm{Der}(L)$ taking an element $x \in L$ to the derivation x^* defined by

$$yx^* = [y,x] \qquad (y \in L).$$

We call * the <u>adjoint representation</u> of L. Derivations of the for x^* for $x \in L$ are <u>inner</u> derivations. The set of all inner derivations of L is written Inn(L). Since the kernel of * is $\zeta_1(L)$, we have $\mathrm{Inn}(L) \cong L/\zeta_1(L)$.

The group of all automorphisms of L is written Aut(L).

There is a useful connection between automorphisms and derivations. Let δ be a nilpotent derivation, so that $\delta^n = 0$ for some n. Then we may define the <u>exponential</u>

$$\exp(\delta) = 1 + \delta + \frac{1}{2!}\delta^2 + \frac{1}{3!}\delta^3 + \dots \ .$$

Similarly if α is a <u>unipotent</u> automorphism, so that $(\alpha-1)^n = 0$ for some n, we may define the logarithm

$$\log(\alpha) = (\alpha-1) - \tfrac{1}{2}(\alpha-1)^2 + \tfrac{1}{3}(\alpha-1)^3 - \dots \ .$$

<u>Proposition 1.1</u> <u>If</u> δ <u>is a nilpotent derivation of</u> L <u>then</u> $\exp(\delta)$ <u>is a unipotent automorphism; and if</u> α <u>is a unipotent automorphism of L then log</u>(α) <u>is a nilpotent derivation.</u>

<u>Proof</u>: The orthodox proof of the first fact may be found in Jacobson [39] p. 9. A similar proof of the second leads to complicated combinatorial problems. The following less orthodox sketch-proof eliminates combinatorial questions altogether, and is at least amusing.

Associate with any linear map $\beta: L \to L$ 'operators' $\underline{L}$ and $\underline{R}$ 'defined' as follows:

$$[x,y]\underline{L} = [x\beta, y]$$
$$[x,y]\underline{R} = [x, y\beta].$$

($\underline{L}$ and $\underline{R}$ are not necessarily well-defined, but this can be steered around by working at a more formal level.) Since

$$[x,y]\underline{LR} = [x\beta, y\beta] = [x,y]\underline{RL}$$

the operators $\underline{R}$ and $\underline{L}$ commute. Now, if β is a derivation we have

$$[x,y]\beta = [x\beta,y] + [x,y\beta] = [x,y](\underline{L}+\underline{R}).$$

Hence

$$[x,y]\exp(\beta) = [x,y]\exp(\underline{L}+\underline{R})$$
$$= [x,y]\exp(\underline{L})\exp(\underline{R})$$

(since $\underline{L}$ and $\underline{R}$ commute)

$$= [x\exp(\beta), y\exp(\beta)].$$

Thus $\exp(\beta)$ is an automorphism. Conversely, if β is an automorphism then

$$[x,y]\beta = [x\beta,y\beta] = [x,y]\underline{LR},$$

so

$$[x,y]\log(\beta) = [x,y]\log(\underline{LR})$$
$$= [x,y](\log(\underline{L})+\log(\underline{R}))$$
$$= [x\log(\beta),y] + [x, y\log(\beta)].$$

Thus $\log(\beta)$ is a derivation. □

There is a slight generalization of this result, where nilpotence of the derivation is replaced by 'local' nilpotence (see [1] p15) but we won't need it.

<u>Proposition 1.2</u> <u>Let</u> δ <u>be a nilpotent derivation of</u> L. <u>Then a subspace</u> M <u>of</u> L <u>is</u> δ-<u>invariant if and only if it is invariant under</u> $\exp(\delta)$.

<u>Proof</u>: See Hartley [27], or [1] p.17. □

A related, but deeper, result whose proof uses algebraic groups is implicit in Chevalley [14]. Proofs are to be found in Tuck [74] and Towers [73]:

Theorem 1.3 *Let* L *be a finite-dimensional Lie algebra over* k *and let* M *be a subspace of* L *invariant under all automorphisms of* L. *Then* M *is invariant under all derivations of* L. □

A subspace of L invariant under every derivation of L is called a *characteristic ideal* of L. We write H ch L if H is a characteristic ideal. The important property is that if H ch K $\triangleleft$ L then H $\triangleleft$ L. Each of $L^{(n)}$, L^n, $\zeta_n(L)$ is a characteristic ideal.

In any finite-dimensional Lie algebra L there is a unique maximal nilpotent ideal $\nu(L)$, called the *nil radical*, and a unique maximal soluble ideal $\sigma(L)$, called the *radical*.

Lemma 1.4 *If* L *is finite-dimensional and* δ *is a derivation of* L *then* $\sigma(L)\delta \subseteq \nu(L)$. *In particular the nil radical and the radical are characteristic ideals of* L. □

If $\sigma(L) = 0$ then L is *semisimple*. For any finite-dimensional L we have $L/\sigma(L)$ semisimple. A useful test for semisimplicity is to introduce the *Killing form*

$$(x,y) = \text{trace } x^*y^*$$

on L. We then have *Cartan's criterion*:

Theorem 1.5 *A finite-dimensional Lie algebra* L *over* k *is semisimple if and only if the Killing form is nonsingular.* □

Using properties of the Killing form one obtains a more structural description:

Theorem 1.6 *A finite-dimensional Lie algebra over* k *is semisimple if and only if it is a direct sum of simple ideals. The summands are unique, and are the minimal ideals of the Lie algebra.* $\square$

Use of the Killing form also yields:

Proposition 1.7 *Every derivation of a finite-dimensional semisimple Lie algebra is inner.* $\square$

Next we turn to representation theory. First, let L be a finite-dimensional nilpotent Lie algebra and M a (finite-dimensional) L-module. Let $\lambda : L \to k$ be a 1-dimensional representation (so that λ is linear and vanishes on L^2). We define the *weight-space*

$$M_\lambda = \{m \in M : m(x^* - \lambda(x))^n = 0 \text{ for some } n\}.$$

Theorem 1.8 *If* M *is a module for the finite-dimensional nilpotent Lie algebra* L *over* k, *then* $M = \bigoplus_\lambda M_\lambda$ *as* λ *runs through all 1-dimensional representations of* L. *Each* M_λ *is a submodule.* $\square$

It follows that a basis for M may be chosen relative to which each $x \in L$ acts as a matrix of the form

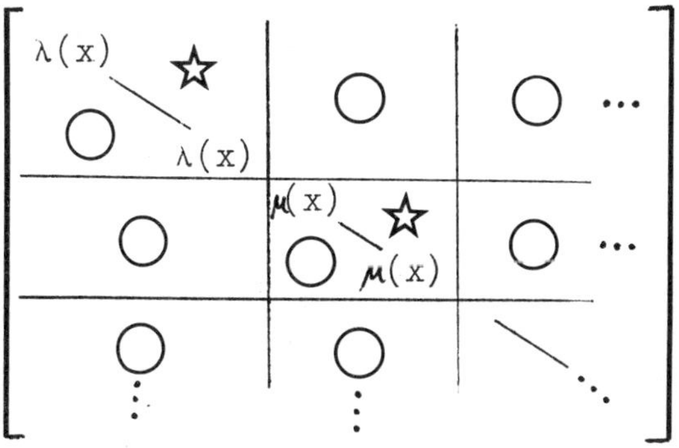

where $\lambda, \mu, \ldots$ are the 1-dimensional representations. If $M_\lambda \neq 0$ we say that λ is a weight of M. Evidently weight-spaces are preserved by intersections with submodules or by taking quotients.

When L is soluble, we can say less about modules. The key result is Lie's theorem:

Theorem 1.9 An irreducible module for a finite-dimensional soluble Lie algebra over k has dimension 1. □

It follows that given any module M a basis may be chosen with respect to which each $x \in L$ acts as a matrix of the form

$$\begin{bmatrix} \lambda_1(x) & & & \star \\ & \lambda_2(x) & & \\ & & \ddots & \\ \bigcirc & & & \lambda_s(x) \end{bmatrix}$$

The simple algebras are classified by using the representation theory of nilpotent algebras. Here the crucial idea is that of a Cartan subalgebra. A subalgebra H of a finite-dimensional Lie algebra L is a Cartan subalgebra if

H is nilpotent (1)

H is self-idealizing: $H = I_L(H)$. (2)

In order to construct Cartan subalgebras we need another concept. For $x \in L$ we can think of L as an $\langle x \rangle$-module with adjoint action, and decompose into weight-spaces $L_0, L_{\lambda_1}, \ldots, L_{\lambda_r}$. Since $[x,x] = 0$ we have $L_0 \neq 0$. We say x is regular if dim L_0 is minimal (as x ranges over all elements of L).

The corresponding L_0 is the <u>null-component</u> of x.

<u>Theorem 1.10</u> <u>If x is regular then its null-component L_0 is a Cartan subalgebra of L.</u> □

The proof of this fact uses:

<u>Lemma 1.11</u> (Engel's theorem) <u>If L is a Lie algebra of finite dimension over k, then L is nilpotent if and only if every inner derivation x^* $(x \in L)$ is nilpotent.</u> □

Let L be a finite-dimensional Lie algebra with a Cartan subalgebra H, and decompose L into weight-spaces L_λ. We have $L_0 = H$, so that

$$L = H \oplus \bigoplus_{\lambda \neq 0} L_\lambda .$$

This is the <u>Cartan decomposition</u> of L. It can be shown that

$$[L_\lambda, L_\mu] \subseteq L_{\lambda+\mu}. \qquad (3)$$

The classification of simple Lie algebras over k is based on a detailed analysis of this decomposition, which reveals a natural geometric structure on the set of weights. The possible geometric configurations are classified, and then the Lie algebras. Although this is the most important part of the classical theory we shall not need it for future work: we need instead the more 'general' theorems.

Modules for semisimple algebras may also be classified. Again we do not need the detailed results, but we do need the general theorem of complete reducibility:

<u>Theorem 1.12</u> <u>Every module for a finite-dimensional semisim-</u>

ple Lie algebra is a direct sum of irreducible submodules. □

Although Cartan subalgebras of Lie algebras are not unique, they are unique 'up to automorphisms'. To make this precise we follow Winter [78] and Humphreys [38] and introduce a very useful group of automorphisms. Let L be finite-dimensional over k. We say that $x \in L$ is strongly nilpotent if there exists an element $y \in L$ and a non-zero weight λ for L as adjoint <y>-module such that $x \in L_\lambda$. It follows from (3) that x^* is nilpotent, so we may form the automorphism $\exp(x^*)$. We define a group

$$\mathcal{E}(L)$$

to be the group of all automorphisms generated by $\exp(x^*)$ for strongly nilpotent x. It has the following 'functorial' properties:

Theorem 1.13 (a) If H is a subalgebra of L then every element of $\mathcal{E}(H)$ extends to an element of $\mathcal{E}(L)$.

(b) An epimorphism $\phi: L \to L'$ induces an epimorphism $\mathcal{E}(L) \to \mathcal{E}(L')$. □

We call these the extension and lifting properties of $\mathcal{E}$.

Theorem 1.14 Let L be a finite-dimensional Lie algebra, with Cartan subalgebras H_1 and H_2. Then there exists $\alpha \in \mathcal{E}(L)$ such that $H_1^\alpha = H_2$. □

(Note that we are using the convention whereby automorphisms are written as superscripts to distinguish them. So H_1^α is the image of H_1 under α.) We say that H_1 and H_2 are

$\mathcal{E}(L)$-conjugate.

Another conjugacy theorem relates to the Borel subalgebras of L, defined to be the maximal soluble subalgebras. Namely

Theorem 1.15 *Let L be a finite-dimensional Lie algebra over k with Borel subalgebras B_1 and B_2. Then there exists $\alpha \in \mathcal{E}(L)$ such that $B_1^{\alpha} = B_2$.* □

Cartan subalgebras may be characterised in a formation-theoretic manner. A class $\mathfrak{F}$ of Lie algebras, closed under isomorphism, is called a formation if

When $L \in \mathfrak{F}$ and $H \triangleleft L$ then $L/H \in \mathfrak{F}$, (4)

If $H, K \triangleleft L$ and $L/H, L/K \in \mathfrak{F}$, then $L/(H \cap K) \in \mathfrak{F}$. (5)

A subalgebra P of L is an $\mathfrak{F}$-projector if

$P \in \mathfrak{F}$, (6)

Whenever H,K are subalgebras of L such that $P \leq H \triangleright K$ and $H/K \in \mathfrak{F}$, then $H = K+P$. (7)

Theorem 1.16 *Let $\mathfrak{N}$ be the formation of nilpotent Lie algebras. Then the $\mathfrak{N}$-projectors of any finite-dimensional Lie algebra L are precisely the Cartan subalgebras. If $\mathfrak{S}$ denotes the formation of soluble Lie algebras, then the $\mathfrak{S}$-projectors are the Borel subalgebras.* □

For proofs, see Barnes and Gastineau-Hills [6].

Corollary 1.17 *Let H be a subalgebra of L, $I \triangleleft L$. If H is a Cartan subalgebra of L then $(H+I)/I$ is a Cartan subalgebra of L/I. If H is a Borel subalgebra of L then $(H+I)/I$ is a Borel subalgebra of L/I.* □

To complete the trinity we introduce the Levi subalgebras of L: semisimple subalgebras complementing the radical.

Theorem 1.18 *Let L be a finite-dimensional Lie algebra over* $\mathfrak{k}$. *Then:*

(a) *Every maximal semisimple subalgebra is a Levi subalgebra, and conversely,*

(b) *If* Λ_1 *and* Λ_2 *are Levi subalgebras of L then there exists* $\alpha \in \mathcal{E}(L)$ *such that* $\Lambda_1^{\alpha} = \Lambda_2$.

Proof: For the first part, see Jacobson [39]. The second is usually proved with α replaced by $\exp(x^*)$ where $x \in \dot{\mathcal{V}}(L)$. To extract $\mathcal{E}(L)$-conjugacy from this is quite easy: we relegate the proof to chapter 11. □

2 Résumé: Algebraic groups

The object of this chapter is to outline the basic facts about algebraic groups needed for applications in later chapters. We adopt an elementary and 'old-fashioned' viewpoint, making no attempt at maximum generality. This approach has the advantage of accessibility, and is sufficient for our immediate purposes. For more information, a standard work such as Borel [9] or Demazure and Gabriel [16] should be consulted.

We begin with some elementary affine algebraic geometry. Let k be an algebraically closed field, and write k^n for the vector space of n-tuples $(x_1,\ldots,x_n) = x$ over k. Let

$$R = k[t_1,\ldots,t_n]$$

be the ring of polynomials over k in variables $t_1,\ldots,t_n$. Given any polynomial $p \in R$ there is a map $k^n \to k$ which sends $x = (x_1,\ldots,x_n)$ to $p(x) = p(x_1,\ldots,x_n)$, and we call this the <u>polynomial function</u> defined by p. Since k is infinite we may identify polynomials with the corresponding polynomial functions. More generally, if V is any n-dimensional vector space over k we may define a polynomial function on V to be a map of the form ip where $i: V \to k^n$ is a linear isomorphism and $p: k^n \to k$ is a polynomial map. It is clear that this yields the same set of maps, independently of the choice of i. We may similarly define polynomial maps $k^n \to k^m$ to be maps sending x to $(p_1(x),\ldots,p_m(x))$ where $p_1,\ldots,p_m$ are polynomials, and extend this to polynomial maps $V \to W$ where V and W are vector spaces over k of dimensions n, m respectively. Once more a change of basis makes no difference to the set of maps.

An <u>affine algebraic variety</u> in a vector space V is the set of points $x \in V$ satisfying a set of polynomial equations $p_i(x) = 0$. If P is any set of polynomials we denote the corresponding variety by

$$V(P) = \{x \in V: p(x) = 0 \text{ for all } p \in P\}.$$

It is easy to verify the following formulae:

$$V(P_1) \cup V(P_2) = V(P_1P_2) \tag{1}$$

$$\bigcap_{i \in I} V(P_i) = V(\bigcup_{i \in I} P_i) \tag{2}$$

$$V(\{1\}) = \emptyset \tag{3}$$

$$V(\{0\}) = V. \tag{4}$$

These imply that the sets V(P) for all possible subsets P of R form the closed sets of a topology on V, called the <u>Zariski topology</u>.

Further, if $P_1 \subseteq P_2$ then $V(P_1) \supseteq V(P_2)$. If P generates the ideal J in R then $V(P) = V(J)$.

Conversely to any subset $E \subseteq V$ we may associate an ideal

$$I(E) = \{p \in R: p(x) = 0 \text{ for all } x \in E\}.$$

There are corresponding results:

$$\bigcap_{j \in J} I(E_j) = I(\bigcup_{j \in J} E_j) \tag{5}$$

$$I(\emptyset) = R \tag{6}$$

$$I(V) = \{0\} \tag{7}$$

and if $E_1 \subseteq E_2$ then $I(E_1) \supseteq I(E_2)$. If a bar denotes the Zariski closure then $V(I(E)) = \bar{E}$ and $I(\bar{E}) = I(E)$. The maps V and I define a bijection between the varieties contained in V and a certain set of ideals of the ring of polynomial maps $V \to k$. In fact the ideals which arise are precisely those ideals J such that if $p^r \in J$ for some r then $p \in J$. This is essentially the strong form of the Hilbert Nullstellensatz.

A topological space is said to be T_1 if singletons are closed sets. It is noetherian if it satisfies the ascending chain condition for open sets, or equivalently the descending chain condition for closed sets.

Proposition 2.1 *Let V be a finite-dimensional vector space over k, equipped with the Zariski topology. Then V is T_1 and noetherian.*

Proof: Obviously V is T_1. There is an order-reversing bijection between closed sets in V and certain ideals in R. Since R has ascending chain condition for ideals (Hilbert basis theorem) it follows that V has descending chain condition for closed sets. □

It follows that the subspace topology on any subset X of V is also T_1 and noetherian. We call this the *Zariski topology* on X. In particular every algebraic variety carries a Zariski topology. It should also be noted that any noetherian topological space is compact (use the finite intersection property for closed sets).

A non-empty topological space is *irreducible* if every pair of non-empty open sets have non-empty intersection (the antithesis of the Hausdorff property). A variety is *irreducible* if it is irreducible as topological space with the Zariski topology. This happens if and only if the corresponding ideal is prime. Every variety is a union of finitely many irreducible varieties.

If V, W are vector spaces over k containing varieties X, Y respectively, then a <u>morphism</u> $\alpha : X \rightarrow Y$ is the restriction to X of a polynomial map $p : V \rightarrow W$ such that $p(X) \subseteq Y$.

For example, let V be a 2-dimensional space over k with coordinate system (x,y). The set of zeros of the polynomial equation

$$x^2 - y^3 = 0$$

is a variety X. (In the case $k = \mathbb{R}$ it is a semicubical parabola as illustrated.)

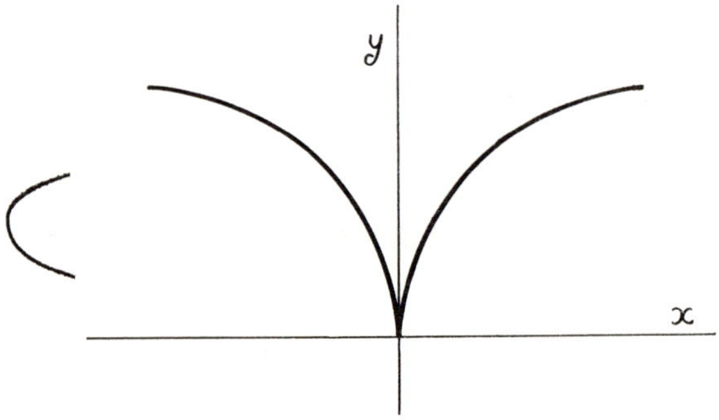

Let $W = k$. Define a map $\alpha : W \rightarrow X$ by $\alpha(w) = (w^3, w^2)$. Then α is a morphism. Note that although α is a bijection its inverse is not a morphism (since it is not the restriction of a polynomial map).

If V is an n-dimensional vector space over k then the set $gl(V)$ of all linear maps $V \rightarrow V$ is a vector space of dimension n^2. The set of invertible maps $GL(V) \subseteq gl(V)$ is a group. If H is a subgroup of $GL(V)$ of the form

$$H = GL(V) \cap X$$

where X is a variety in $gl(V) \cong k^{n^2}$, then we call H a **linear** <u>algebraic group</u>. We do not consider other kinds of algebraic groups, and will drop the adjective 'linear'.

For example, GL(V) is an algebraic group. The special linear group SL(V) is obtained by taking X to be the set of linear maps of determinant 1. Clearly the map $x \to \det(x)-1$ is polynomial, for if we represent x by a matrix (x_{ij}) the determinant is a polynomial function of the x_{ij}.

A minor discomfort with this definition is that GL(V) is not itself an affine algebraic variety. It is, on the contrary, an <u>open</u> subset of gl(V), inasmuch as it is the complement of the maps of determinant zero. This of course happens because we have chosen too restricted a definition of 'variety'. A more general definition, though more useful in the long run (indeed essential), introduces complications which are unnecessary here.

The important instances of algebraic groups, for us, arise as follows:

<u>Theorem 2.2</u> <u>Let A be a finite-dimensional algebra over k. Then the group Aut(A) of all automorphisms of A is algebraic.</u>

<u>Proof</u>: Take a basis $\{a_1, \ldots, a_n\}$ for A and define structure constants $N_{ijk} \in k$ by

$$a_i a_j = \Sigma_k N_{ijk} a_k.$$

Let $\alpha: A \to A$ be defined by

$$a_i \alpha = \Sigma_j \alpha_{ij} a_j.$$

Then the condition that α be an automorphism translates into a system of polynomial equations for the α_{ij}. Since moreover α must be invertible, Aut(A) is the intersection of GL(A) and a certain variety, so is an algebraic group. □

Since an algebraic group is a subset of gl(V) for some V it carries the Zariski topology. This is what makes the algebraic group concept useful in the sequel, for it allows us to invoke topological arguments.

Suppose $H \subseteq gl(V)$ and $K \subseteq gl(W)$ are algebraic groups. A map $\alpha: H \to K$ is a <u>morphism</u> (of algebraic groups) if:

α is a group homomorphism (8)

α is the restriction to H of a polynomial map $p: V \to W$ such that $p(H) \subseteq K$. (9)

Any morphism from one variety to another is continuous in the Zariski topology. However, it need not be open or closed. Thus the morphism $\alpha: k^2 \to k^2$ defined by

$$\alpha(x,y) = (xy,y)$$

has as image $k \times (k \setminus \{0\}) \cup \{(0,0)\}$ which is not open and not closed. (It is, however, the union of an open set and a closed set.) Morphisms of algebraic groups are better behaved. The reasons for this are as follows. In k^n with Zariski topology (one can be more general) a subset X is said to be <u>locally closed</u> if it satisfies any of the following three equivalent conditions:

X is open in its closure $\bar{X}$,

$X = Y \cap Z$ where Y is open and Z is closed,

Each $x \in X$ has a neighbourhood N_x in k^n such that $X \cap N_x$ is closed in N_x.

A subset of k^n is constructible if it is the union of a finite collection of locally closed subsets. Equivalently it belongs to the Boolean algebra generated by the closed subsets of k^n.

Proposition 2.3 *If $\alpha : X \to Y$ is a morphism of varieties, and if K is a constructible subset of X, then $\alpha(K)$ is a constructible subset of Y.* □

In particular the image of a variety under a morphism is constructible. If G is an algebraic group, H is a subgroup of G, and H is constructible, then it may be shown that H is closed. Combining this with the above proposition we get:

Theorem 2.4 *If $\alpha : G \to G'$ is a morphism of algebraic groups and H is an algebraic subgroup of G, then $\alpha(H)$ is an algebraic subgroup of G'.* □

(An algebraic subgroup is one which is Zariski-closed.) It is crucial in 2.4 to have the blanket assumption that k is algebraically closed. Thus, if G is the multiplicative group of nonzero real numbers $GL_1(\mathbb{R})$ and $\alpha : G \to G$ is defined by $\alpha(x) = x^2$ it is manifest that α is a morphism; but $\alpha(G)$ is the group of positive reals and this is not Zariski-closed.

It is important to note that an algebraic group is not a topological group with respect to the Zariski topology. The group operation $G \times G \to G$ is continuous, not with respect to the product topology on $G \times G$, but with respect to the Zariski topology (as a subset of some $gl(V) \times gl(V)$).

Proposition 2.5 *Let G be an algebraic group and H a subgroup of G. Then $\bar{H}$ is a subgroup of G, and is the smallest algebraic subgroup of G containing H.* □

It is not true that a closed subset of an algebraic group generates a closed subgroup. In fact this is not even true for topological groups. The set of all 2×2 matrices over $\mathbb{C}$ of the form

$$\begin{pmatrix} 0 & z \\ 0 & 0 \end{pmatrix} \qquad (z \in \mathbb{C})$$

is an algebraic group isomorphic to the additive group $\mathbb{C}^+$, and is a topological group in the usual (metric) topology. We may identify each such matrix with the element $z \in \mathbb{C}^+$. Now the closed set $\{1, \sqrt{2}\}$ generates a (metrically) dense subgroup D of $\mathbb{R}^+$, and since this is not the whole of $\mathbb{R}^+$, D is not closed. But the Zariski topology is coarser than the metric topology, because polynomials are metrically continuous, so D is not algebraic either.

Better behaviour occurs under connectivity restrictions. Recall that a topological space is *connected* if it is not the disjoint union of two non-empty open (hence closed) subsets. As usual we use the Zariski topology to define connectedness of a variety. Every variety is the disjoint union of its connected components. If G is an algebraic group we write G^0 for the connected component of the identity element.

Theorem 2.6 *Let G be an algebraic group. Then*

(a) *G^0 is an algebraic subgroup of G.*

(b) G^o _is a normal subgroup of_ G _and_ G/G^o _is finite_.

(c) _The cosets of_ G^o _are the connected, and also the irreducible, components of_ G. □

Proposition 2.7 _Let_ G _be an algebraic group, and let_ H,K _be closed connected subgroups. Then the subgroup_ <H,K> _generated by_ H _and_ K _is a closed connected subgroup_. □

Obviously proposition 2.7 extends inductively to any finite set of closed connected subgroups. Actually, the finiteness of the set is unnecessary. To see this we must examine the concept of 'dimension' for algebraic varieties.

Consider a noetherian topological space X. We define the _combinatorial dimension_ of X to be the supremum of the lengths c of chains

$$X_0 \subsetneq X_1 \subsetneq X_2 \subsetneq \cdots \subsetneq X_c$$

of distinct irreducible closed sets X_i in X. We denote it by

$$\dim X.$$

It is either a positive integer, ∞, or (conventionally) $-\infty$ when $X = \emptyset$. If X is the union of finitely many irreducible sets Y_i then

$$\dim X = \sup_i \dim Y_i.$$

When X is an irreducible algebraic variety, dim X is equal to the transcendence degree of a certain field associated with X and is finite. Hence every variety has finite dimension. In an irreducible variety of dimension d every chain of closed irreducible subsets can be refined to one of length exactly d. Now a connected algebraic group is irreducible by 2.6(c), and we have:

<u>Theorem 2.8</u> *Let G,H be connected algebraic groups with H a subgroup of G. Then*

(a) $\dim H \leq \dim G$,

(b) *If* $\dim H = \dim G$ *then* $H = G$. □

Even if G is not connected we have $\dim G = \dim G^o$, and if H is a subgroup of G then $H^o \subseteq G^o$. Hence we obtain:

<u>Theorem 2.9</u> *Every algebraic group satisfies the ascending chain condition for closed connected subgroups.* □

Using this in conjunction with proposition 2.7 and the obvious induction argument, we obtain:

<u>Theorem 2.10</u> *If* $\{H_i\}_{i \in I}$ *is a family of closed connected subgroups of an algebraic group G, then the subgroup* $\langle H_i : i \in I \rangle$ *which they generate is a closed connected subgroup of G. Further, this subgroup is equal to* $\langle H_i : i \in I_o \rangle$ *for some finite subset* $I_o \subseteq I$. □

This result is useful in the following context. If L is a finite-dimensional Lie algebra over k and $\delta : L \to L$ is a derivation which is nilpotent, so that $\delta^c = 0$ for some c, then we may define the exponential $\exp(\delta)$ and this is an automorphism of L. A 1-*parameter subgroup* of Aut(L) is a subgroup of the form

$$E_\delta = \{\exp(\lambda\delta) : \lambda \in k\}.$$

Now the map $e : k^+ \to E_\delta$ defined by $e(\lambda) = \exp(\lambda\delta)$ is a morphism of algebraic groups (thinking of k^+ as the group of matrices $\begin{pmatrix}0 & \lambda \\ 0 & 0\end{pmatrix}$) because $\exp(\lambda\delta)$ is a polynomial in $\lambda\delta$. Now k^+ is a connected algebraic group: hence E_δ is connected (by the obvious topological argument) and closed (by theorem 2.4), so is a closed connected subgroup of Aut(L). Therefore we may invoke theorem 2.10 to obtain:

Theorem 2.11 If L is a finite-dimensional Lie algebra over k and G is a subgroup of Aut(L) generated by 1-parameter subgroups, then G is a connected algebraic group. □

Corollary 2.12 The group $\mathcal{E}(L)$ is a connected algebraic group. □

Finally, let $G \subseteq GL(V)$ be an algebraic group, and let W be a subspace of V. Define the normalizer (or stabilizer)

$$N_G(W) = \{\alpha \in G : \alpha(W) \subseteq W\}.$$

Proposition 2.13 With the above notation, $N_G(W)$ is an algebraic subgroup of G. □

A more general result, where V is replaced by a variety on which G acts 'morphically', is proved in Borel [9] p.97.

3 Locally and ideally finite Lie algebras

Let $\mathfrak{X}$ be a class of Lie algebras. A Lie algebra L is said to be <u>locally-$\mathfrak{X}$</u> if every finite subset of L is contained in a subalgebra which belongs to $\mathfrak{X}$. If, as is usually the case, $\mathfrak{X}$ is subalgebra-closed, then equivalently every finite subset generates a subalgebra belonging to $\mathfrak{X}$. When $\mathfrak{X}$ is respectively the class of nilpotent, soluble, or finite-dimensional algebras we say that L is <u>locally nilpotent</u>, <u>locally soluble</u>, or <u>locally finite</u>.

The canonical example of a locally finite Lie algebra is obtained as follows. Let V be any vector space. Recall that a linear map $\alpha: V \to V$ has <u>finite rank</u> if its image $V\alpha$ has finite dimension. The set $\mathfrak{F}(V)$ of all such α is an associative algebra, so becomes a Lie algebra under commutation

$$[\alpha, \beta] = \alpha\beta - \beta\alpha.$$

<u>Proposition 3.1</u> <u>The Lie algebra $\mathfrak{F}(V)$ is locally finite.</u>

<u>Proof</u>: Let $\alpha_1, \ldots, \alpha_n \in \mathfrak{F}(V)$. Let

$$W = \Sigma_{i=1}^{n} \operatorname{im}(\alpha_i),$$

$$K = \bigcap_{i=1}^{n} \ker(\alpha_i).$$

Then both W and V/K have finite dimension. The set

$$H = \{\alpha \in F(V) : V\alpha \subseteq W, K\alpha = 0\}$$

is a subalgebra of finite dimension containing $\alpha_1, \ldots, \alpha_n$. □

As a variation, one may totally order a basis $\{v_i\}$ of V and consider only those linear maps of finite rank which map each v_i to a linear combination of larger v_j. This yields locally nilpotent Lie algebras of 'locally zero-triangular' maps, analogous to groups constructed by McLain [48]. If we allow v_j equal to v_i as well as larger, we get locally soluble algebras of 'locally triangular' maps. Many ideas of linear algebra in finite dimensions can be generalized to finite rank transformations. In particular if $\alpha \in \underset{\sim}{F}(V)$ we can define the trace of α to be the trace of its restriction to any finite dimensional space containing $V\alpha$. The trace is well-defined, satisfies the usual linearity and commutation properties, and is a Lie homomorphism

$$\text{trace: } \underset{\sim}{F}(V) \rightarrow k .$$

The set of maps of trace zero is a simple ideal $\underset{\sim}{T}(V)$ of $\underset{\sim}{F}(V)$, and is in fact the only non-trivial ideal of $\underset{\sim}{F}(V)$ (see, for example, [61]). Unlike finite dimensions, $\underset{\sim}{T}(V)$ does not have a direct complement in $\underset{\sim}{F}(V)$ because scalar multiples of the identity map do not have finite rank when dim V is infinite.

A local system for a Lie algebra L is a collection $\{L_i\}_{i \in I}$ of subalgebras of L which generate L and have the property that whenever $i, j \in I$ there exists $k \in I$ such that $\langle L_i, L_j \rangle \leq L_k$. It follows that

$$L = \bigcup_{i \in I} L_i .$$

We can define a partial order $\leq$ on I by setting $i \leq j$ if and only if $L_i \leq L_j$, and this makes I into a directed set. We call $\leq$ the inclusion ordering on I.

If L has a local system of $\mathfrak{X}$-subalgebras, for some class $\mathfrak{X}$, then obviously L is a locally-$\mathfrak{X}$-algebra. At least when $\mathfrak{X}$ is subalgebra-closed, the converse is true.

There is an interesting hierarchy of subclasses of the class of locally finite Lie algebras, obtained by placing more or less stringent restrictions on a local system. We recall that a subalgebra H of a Lie algebra L is <u>ascendant</u> if there is an ordinal σ and a series of subalgebras $\{L_\alpha\}_{\alpha\leq\sigma}$ such that

$$L_0 = H, \quad L_\sigma = L,$$

$$L_\alpha \triangleleft L_{\alpha+1} \text{ for all } \alpha < \sigma,$$

$$L_\lambda = \bigcup_{\alpha<\lambda} L_\alpha \text{ for all limit ordinals } \lambda \leq \sigma.$$

If σ is finite then H is a <u>subideal</u> of L. (There is also a more general concept of a <u>serial</u> subalgebra, where the ordinal σ is replaced by an arbitrary totally ordered set, dealt with in [1] pp.27, 258.)

We say that a Lie algebra L is:

<u>ideally finite</u> if it has a local system of finite-dimensional ideals;

<u>subideally finite</u> if it has a local system of finite-dimensional subideals,

<u>ascendantly finite</u> if it has a local system of finite-dimensional ascendant subalgebras,

<u>serially finite</u> if it has a local system of finite-dimensional serial subalgebras.

Ascendantly finite algebras were called $\acute{N}\mathfrak{F}$-algebras in [54], serially finite algebras are the 'neoclassical' algebras or $\mathfrak{h}$-algebras of [1], and ideally finite algebras are called $\bar{\mathfrak{F}}$-algebras in [57, 58]. In these notes we shall concentrate on ideally finite algebras, the most restricted class in the hierarchy, and make only brief references to the others. By considering locally nilpotent algebras and using the results of [1] chapter 6 it is easy to see that these classes, and the class of locally finite Lie algebras, are distinct.

We can characterize ideally finite algebras in several ways:

<u>Theorem 3.2</u> <u>The following properties of a Lie algebra L are equivalent:</u>

(a) L <u>is ideally finite,</u>

(b) <u>Every element of L lies in a finite-dimensional ideal of</u> L,

(c) L <u>is generated by finite-dimensional ideals,</u>

(d) <u>The adjoint map</u> $x^*: L \to L$ <u>has finite rank for every</u> $x \in L$,

(e) <u>The centralizer</u> $C_L(x)$ <u>has finite codimension in</u> L <u>for every</u> $x \in L$.

<u>Proof</u>: Since $C_L(x)$ is the kernel of x^*, (d) and (e) are equivalent. The only non-trivial implication is that (e) implies (b), which is proved as follows. Let $H = h_1, \ldots, h_n$ be a finitely generated subalgebra of L. Then $\bigcap_{i=1}^{n} C_L(h_i)$ is central in H and has finite codimension, so H/Z is finite-dimensional where $Z = \zeta_1(H)$. Choose a vector space complement T to Z in H. Then $[[T,T],T] \subseteq [Z+T,T] \subseteq [T,T]$, so that $\langle T \rangle = T + [T,T]$. Now H is generated by T together with a finite subset $\{z_1, \ldots, z_m\}$ of Z, and hence

$$H = T + [T,T] + \langle z_1 \rangle + \ldots + \langle z_m \rangle$$

which has finite dimension. **Thus** L is locally finite. Now for $x \in L$ we can find a finite-dimensional subalgebra X such that $L = C_L(x)+X$. By [1] lemma 2.2.3 p.45 we have

$$x^L = (x^{C_L(x)})^X = x^X \leq \langle x,X \rangle$$

which is finite-dimensional. Hence (b) holds. □

The analogy with FC-groups is apparent on two levels. If one defines an FC-group to be one in which all conjugacy classes are finite, or equivalently centralizers of elements have finite index, one obtains the obvious analogue of (e). On the other hand, observing that a periodic FC-group is generated by finite normal subgroups, one obtains (c). Thus the distinction between periodicity and non-periodicity vanishes for Lie algebras.

Examples of ideally finite algebras are easy to find. The most obvious are direct sums $\mathrm{Dr}_{i \in I} F_i$ of finite-dimensional algebras F_i. Less trivially, quotients of subalgebras of such direct sums (by virtue of theorem 3.2(b)). There is no reason to expect that the latter decompose into direct sums of finite-dimensional algebras.

Other examples arise as algebras of linear transformations of a vector space V equipped with two families of subspaces $\{V_i\}_{i \in I}$, $\{W_i\}_{i \in I}$. Suppose that the V_i have finite dimension and the W_i finite codimension. Let M be the Lie algebra of all linear maps $V \to V$ which leave every V_i and every W_i invariant. For each $i \in I$ let F_i be the set of $\alpha \in M$ such that

$V\alpha \subseteq V_i$ and $W_i\alpha = 0$. Then F_i is finite-dimensional (since any α is uniquely specified by its restriction to a vector space complement K to W_i as a map $K \to V_i$) and is an ideal of M since if $\alpha \in F_i$, $\beta \in M$, then

$$V(\alpha\beta-\beta\alpha) \subseteq V\alpha.\beta + V\beta.\alpha \subseteq V_i\beta+V_i \subseteq V_i,$$

$$W_i(\alpha\beta-\beta\alpha) \subseteq W_i\alpha.\beta+W_i\beta.\alpha \subseteq 0+W_i\alpha \subseteq 0 .$$

Hence $L = \Sigma_{i \in I} F_i$ is an ideally finite subalgebra of M.

Next we dualize the 'local' concept. If $\mathfrak{X}$ is a class of Lie algebras we say that L is residually-$\mathfrak{X}$ if it has a set $\{K_j\}_{j \in J}$ of ideals such that $\bigcap_{j \in J} K_j = 0$ and each L/K_j belongs to $\mathfrak{X}$. In particular we obtain residually nilpotent, residually soluble, and residually finite Lie algebras by taking $\mathfrak{X}$ to be the class of nilpotent, soluble, or finite-dimensional algebras respectively.

A residual system for L is a set $\{K_j\}_{j \in J}$ of ideals with the properties that $\bigcap_{j \in J} K_j = 0$, and if $i, j \in J$ then there exists $k \in J$ such that $K_k \leq K_i \cap K_j$. In particular L is residually finite if and only if it has a finite residual system, that is, a residual system $\{K_j\}$ with L/K_j finite-dimensional for all j. We can partially order J by reverse inclusion:

$$i \leq j \text{ if and only if } K_i \geq K_j,$$

and it then becomes a directed set.

Residual systems arise in the study of ideally finite algebras because of:

Lemma 3.3 *If L is ideally finite then $L/\zeta_1(L)$ is residually finite.*

Proof: By [62] p.302 it follows that for every finite-dimensional ideal F of L we have $C_L(F)$ of finite codimension. But the intersection of these centralizers, for all such F, is the centre of L. □

The following easy lemma is very useful in the study of ideally finite Lie algebras.

Lemma 3.4 If L is residually finite and X is a finite-dimensional subspace of L, then there exists an ideal $K \triangleleft L$ of finite codimension for which $K \cap X = 0$.

Proof: If $0 \neq x_1 \in X$ then there exists $K_1 \triangleleft L$ of finite codimension such that $x_1 \notin K_1$. Pick $0 \neq x_2 \in X \cap K_1$, unless $X \cap K_1 = 0$, and continue the process to obtain a descending chain of ideals of finite codimension

$$K_1 \geq K_1 \cap K_2 \geq K_1 \cap K_2 \cap K_3 \geq \ldots$$

whose intersections with X decrease strictly. Since X has finite dimension we eventually have $X \cap (K_1 \cap K_2 \cap \ldots \cap K_n) = 0$. □

Subalgebras of direct sums of finite-dimensional algebras are residually finite as well as ideally finite. A partial converse, analogous to a theorem of Hall [26] p.290, may be proved.

Proposition 3.5 The countable-dimensional residually finite Lie algebras are precisely the subalgebras of direct sums of countably many finite-dimensional Lie algebras.

Proof: If L is countable-dimensional residually finite and ideally finite then we can write L as the union of a countable ascending chain $\{L_i\}_{i=1,2,3,...}$ of finite-dimensional ideals. Using lemma 3.4 we can choose ideals K_i of finite codimension such that $K_i \cap L_i = 0$, and it is easy to modify the sequence $\{K_i\}$ to obtain a decreasing sequence. Define $I_{i+1} = L_i + K_{i+1}$ for $i = 1,2,3,...$. Then I_{i+1} is an ideal of finite codimension. Further,

$$L_{i+1} \cap I_{i+1} = L_{i+1} \cap (L_i + K_{i+1}) = L_i + (L_{i+1} \cap K_{i+1}) = L_i$$

and it follows that

$$\bigcap_{i=1}^{\infty} I_{i+1} = 0.$$

There is an obvious monomorphism from L into the Cartesian sum of the L/I_i given by

$$x \to (I_2 + x,\ I_3 + x, ...)$$

but since every $x \in L$ belongs to all but a finite number of the I_i the image of this map is contained in the <u>direct</u> sum of the L/I_i. □

Now by Ado's theorem every finite-dimensional Lie algebra embeds in $gl_n(k)$ for some n. It follows easily that the countable-dimensional residually finite ideally finite Lie algebras are precisely the subalgebras of

$$gl_1(k) \oplus gl_2(k) \oplus gl_3(k) \oplus \cdots \quad .$$

It is perhaps unnecessary to add that this characterisation does not seem to be helpful as regards structural questions about ideally finite Lie algebras!

In the following pages we shall often work with ideally

finite Lie algebras which are locally nilpotent or locally soluble. It should be pointed out that these have a very special structure. Recall that a Lie algebra is <u>hypercentral</u> of <u>height</u> σ if $L = \zeta_\sigma(L)$ and σ is the least ordinal with this property. Hypercentral algebras are always locally nilpotent, but not conversely. To obtain an analogous concept related to local solubility we define the <u>paracentre</u> $\pi_1(L)$ of a Lie algebra L to be the subalgebra generated by its ideals of dimension 1, and define the <u>upper paracentral series</u> $\pi_\alpha(L)$ for ordinals α in the usual manner: $\pi_{\alpha+1}(L)/\pi_\alpha(L) = \pi_1(L/\pi_\alpha(L))$ with unions at limit ordinals. Say that L is <u>hyperparacentral</u> of <u>paraheight</u> σ if $L = \pi_\sigma(L)$ and σ is the least such ordinal. Finally say that L is <u>hypercyclic</u> if it has an ascending series of ideals $(L_\alpha)_{\alpha \leq \sigma}$ such that $\dim L_{\alpha+1}/L_\alpha = 1$ for all $\alpha < \sigma$. Classes of Lie algebras related to these have been studied by Brazier [8].

<u>Theorem 3.6</u> <u>A locally nilpotent ideally finite Lie algebra is hypercentral of height $\leq \omega$. A locally soluble ideally finite Lie algebra is hyperparacentral of paraheight $\leq \omega$.</u>

<u>Proof</u>: Let L be locally nilpotent and ideally finite, F a finite-dimensional ideal of L. Then if $d = \dim F$ we have $F \leq \zeta_d(L)$ by [1] lemma 1.6 p.137. Hence $L \leq \zeta_\omega(L)$.

Now suppose L is locally soluble ideally finite, with F a finite-dimensional ideal. Then F is an $L/C_L(F)$-module under the adjoint action, and $L/C_L(F)$ is a finite-dimensional soluble

Lie algebra. By Lie's theorem F contains a 1-dimensional $L/C_L(F)$-submodule, which clearly lies inside $\pi_1(L)$. Hence if d = dim F we have inductively $F \leq \pi_d(L)$, and $L \leq \pi_\omega(L)$. □

Since hyperparacentral Lie algebras are obviously hypercyclic it follows in particular that locally soluble ideally finite Lie algebras are hypercyclic. In the countable-dimensional case this is a theorem of Vasilesçu [75].

4 Radicals and semisimplicity

From now on, L will unless otherwise stated denote an ideally finite Lie algebra over k. We let $\{F_i\}_{i \in I}$ be the set of all finite-dimensional ideals of L.

It is very easy to obtain suitable radicals for L. We define

$$\nu(L) = \Sigma_{i \in I} \, \nu(F_i),$$
$$\sigma(L) = \Sigma_{i \in I} \, \sigma(F_i).$$

Since $\nu(F_i)$ and $\sigma(F_i)$ are characteristic ideals of F_i they are ideals of L.

<u>Proposition 4.1</u> <u>If</u> L <u>is ideally finite over</u> k, <u>then</u> $\nu(L)$ <u>is the unique maximal locally nilpotent ideal of</u> L, <u>and</u> $\sigma(L)$ <u>is the unique maximal locally soluble ideal of</u> L.

<u>Proof</u>: Clearly $\nu(L)$ is a locally nilpotent ideal. If N is any locally nilpotent ideal of L, then $N \cap F_i$ is a nilpotent ideal of F_i, so is contained in $\nu(F_i)$. Since N is the union of all $N \cap F_i$ it follows that $N \leq \nu(L)$. The proof for $\sigma(L)$ is similar. □

A derivation δ of L is <u>locally finite</u> if every finite subset of L is contained in a finite-dimensional δ-invariant subspace.

<u>Proposition 4.2</u> <u>Let</u> L <u>be ideally finite over</u> k, <u>and let</u> δ <u>be a locally finite derivation of</u> L. <u>Then</u> $\sigma(L)\delta \subseteq \nu(L)$.

<u>Proof</u>: Let $x \in \sigma(L)$. There exists a finite-dimensional

δ-invariant subspace X containing x. We have $X \subseteq F_i$ for some $i \in I$, so the ideal closure X^L is finite-dimensional, hence equal to some F_j. Since $X^L = \Sigma_{n=0}^{\infty} [X,_n L]$ an easy induction shows that X^L is also δ-invariant. Hence $\delta' = \delta|_{F_j}$ is a derivation of F_j. Now $x \in \sigma(L) \cap F_j \leq \sigma(F_j)$, hence $x\delta = x\delta' \in \nu(F_j)$. Therefore $\sigma(L)\delta \subseteq \nu(L)$ as claimed. □

Corollary 4.3 If L is locally soluble ideally finite over k then L^2 is locally nilpotent. □

Corollary 4.4 If L is ideally finite over k then

$$[\sigma(L), L] \leq \nu(L).$$ □

Let us say that an ideal R of L is relatively characteristic if, whenever K is ideally finite and $L \lhd K$, we have $R \lhd K$.

Corollary 4.5 If L is ideally finite then $\nu(L)$ and $\sigma(L)$ are relatively characteristic ideals of L. □

Proposition 4.6 Let L be ideally finite over k, and let H be an ideal of L. Then $\sigma(H) = \sigma(L) \cap H$, $\nu(H) = \nu(L) \cap H$.

Proof: By corollary 4.5, $\sigma(H)$ is a locally soluble ideal of L, so $\sigma(H) \leq \sigma(L) \cap H$. The reverse inclusion is obvious. The proof for $\nu(L)$ is similar. □

If $\sigma(L) = 0$ we say that L is semisimple.

Theorem 4.7 If L is ideally finite over k then $L/\sigma(L)$ is semisimple.

Proof: Let $R/\sigma(L) = \sigma(L/\sigma(L))$. Then $R \triangleleft L$. By considering a local system of finite-dimensional ideals for R it is clear that R is locally soluble, hence $R \leq \sigma(L)$. Therefore $R = \sigma(L)$ as required. □

We can classify the semisimple ideally finite algebras completely:

Theorem 4.8 An ideally finite Lie algebra L over k is semisimple if and only if L is a direct sum of finite-dimensional simple ideals. The summands are unique and are the minimal ideals of L.

Proof: If $\sigma(L) = 0$ then $\sigma(F_i) = 0$ for all i, hence L is a sum of finite-dimensional semisimple ideals. Each of these is a direct sum of finite-dimensional simple ideals S_j. Since $S_j^2 = S_j$ and $S_j \triangleleft F_i \triangleleft L$ it follows (Schenkman [52], [1] p.11) that $S_j \triangleleft L$. Hence L is a sum of finite-dimensional simple ideals. It is well known and easy to prove (e.g. [1] p.263) that this implies that L is a direct sum of finite-dimensional simple ideals. The summands are clearly minimal ideals of L. Uniqueness follows from the next lemma. □

Lemma 4.9 Let $L = \bigoplus_{m \in M} S_m$ where the S_m are finite-dimensional simple. If H is an ideal of L then there is a subset M' of M such that $H = \bigoplus_{m \in M'} S_m$.

Proof: See Vasilesçu [75], or [1] p.264. □

<u>Corollary 4.10</u> <u>Every ideal and every quotient of a semisimple ideally finite algebra over k is semisimple.</u> □

On any ideally finite Lie algebra L we may define the <u>Killing form</u> $(x,y) = \text{trace } x^*y^*$, because x^* and y^* have finite rank. It is left to the reader to verify the usual properties of the Killing form, and to show that Cartan's criterion for semisimplicity generalizes satisfactorily. (We do not use this result in the sequel.)

Since all finite-dimensional simple algebras over k are known, theorem 4.8 effectively reduces problems on the structure of semisimple ideally finite algebras to finite dimensions. For this reason the theory of semisimple ideally finite algebras is 'trivial', and all the interest centres on non-semisimple algebras.

5 The Frattini subalgebra

The <u>Frattini subalgebra</u> of a Lie algebra L is defined to be the intersection of the maximal subalgebras of L, with the convention that this intersection is L if there are no maximal subalgebras. We denote it by F(L). In contrast to the situation prevailing in group theory, F(L) need not be an ideal of L (Barnes [4]), and it is usually necessary to introduce the <u>Frattini ideal</u> $\Phi(L)$, which is the largest ideal contained in F(L). However, there are several known conditions ensuring that F(L) is an ideal of L, as follows:

(i) L finite-dimensional soluble, over any field (see Barnes and Gastineau-Hills [6]).

(ii) L finite-dimensional over any field of characteristic zero (Marshall [47] in the algebraically closed case, Towers [73] or Tuck [74] in general).

(iii) L locally nilpotent, in which case $F(L) = L^2$ (see [1] p.242).

We shall give a somewhat broader criterion, obtained as a generalization of (i), which will in particular apply to ideally finite Lie algebras. Since the results of this section have not been published hitherto we take a fairly general line.

The basic result, whose proof was suggested by Barnes and Gastineau-Hills [6] lemma 3.4, is as follows:

<u>Theorem 5.1</u> <u>Let</u> L <u>be a Lie algebra over any field, having an ascending series</u> $(L_\alpha)_{\alpha \leq \sigma+1}$ <u>for an ordinal</u> σ, <u>with the following properties</u>:

(a) $L_\alpha \triangleleft L$ <u>for all</u> $\alpha \leq \sigma$,

(b) $F(L_{\sigma+1}/L_\sigma) \triangleleft L_{\sigma+1}/L_\sigma$,

(c) $L_{\alpha+1}/L_\alpha$ <u>is abelian for all</u> $\alpha < \sigma$.

<u>Then</u> $F(L)$ <u>is an ideal of</u> L.

<u>Proof</u>: Let $\mathcal{M}$ be the set of all maximal subalgebras of L. If $\mathcal{M} = \emptyset$ then $F(L) = L \triangleleft L$, so we may assume $\mathcal{M} \neq \emptyset$. We shall show that for each $M \in \mathcal{M}$ there exists a subset $\mathcal{M}_M$ of $\mathcal{M}$ such that

$$\bigcap \mathcal{M}_M \triangleleft L.$$

Then

$$F(L) = \bigcap \mathcal{M} = \bigcap_M \bigcap \mathcal{M}_M \triangleleft L$$

as required.

Suppose $M \in \mathcal{M}$, and let α be the largest ordinal $\leq \sigma$ for which $L_\alpha \leq M$. If $\alpha = \sigma$ then $M \geq L_\sigma$, and we let $\mathcal{M}_M$ be the set of all maximal subalgebras of L which contain L_σ. Then $\bigcap \mathcal{M}_M$ is the inverse image in L of $F(L/L_\sigma)$ and this is an ideal by (b).

Otherwise $\alpha < \sigma$. We let $c(M)$ be the largest ideal of L contained in M. Passing to $L/c(M)$ we may assume that $c(M) = 0$, or that M is <u>corefree</u>. Putting

$$A = \frac{L_{\alpha+1} + c(M)}{c(M)}$$

we reach the following situation: M is a corefree maximal subalgebra of a Lie algebra L, having a non-zero abelian ideal A such that $A \nleq M$. By maximality $L = A + M$. Now $M \cap A$ is idealized by M since $A \triangleleft L$, and by A since A is abelian. But this means $M \cap A \triangleleft L$. Since M is corefree, $A \cap M = 0$ and the extension $L = A \dotplus M$ is split. Consider A as an M-module.

If $0 \neq B$ is an M-submodule of A, then B+M is a subalgebra, and maximality implies L = B+M and hence that B = A. Therefore A is irreducible as M-module. Now $C_M(A)$ is easily seen to be an ideal of L, so is zero since M is corefree, so A is a faithful M-module. Hence given $0 \neq m \in M$ there exists $a \in A$ such that $[m,a] \neq 0$. The map $\beta = 1+a^*$ is an automorphism of L, and $m^\beta = m + [m,a] \notin M$, since otherwise we would have $[m,a] \in M \cap A = 0$. Hence $m \notin M^{-\beta}$. Thus if we put

$$\mathcal{M}_M = \{M^{1+a^*} : a \in A\}$$

we have

$$\bigcap \mathcal{M}_M = 0 \triangleleft L.$$

We may lift from L/c(M) to L, and now we have $\bigcap \mathcal{M}_M = c(M)$ which is an ideal of L as required. □

There is an immediate corollary:

<u>Corollary 5.2</u> <u>Let</u> $\mathfrak{X}$ <u>be a class of Lie algebras, closed under quotients. Then</u> $F(L) \triangleleft L$ <u>for all</u> $L \in \mathfrak{X}$ <u>if and only if</u> $F(K) \triangleleft K$ <u>for all</u> $K \in \mathfrak{X}$ <u>such that</u> K <u>contains no non-zero abelian ideals.</u> □

Recall that L is <u>hyperabelian</u> if it has an ascending series of ideals with abelian factors, and that L is $\mathfrak{X}$-by-$\mathfrak{Y}$ if it has an ideal $K \in \mathfrak{X}$ with $L/K \in \mathfrak{Y}$. We then have:

<u>Theorem 5.3</u> <u>Over any field, we have</u> $F(L) \triangleleft L$ <u>if</u> L <u>is</u>

(a) <u>Soluble,</u>

(b) <u>Hyperabelian,</u>

(c) <u>Soluble-by-locally nilpotent,</u>

(d) Hyperabelian-by-locally nilpotent.

Over a field of characteristic zero, $F(L) \triangleleft L$ if L is

(e) Soluble-by-finite-dimensional,

(f) Hyperabelian-by-finite-dimensional,

(g) Ideally finite.

Proof: Everything is straightforward. For (g), note that $\sigma(L)$ has an ascending series with abelian factors whose terms are ideals of L. Modulo $\sigma(L)$, L is semisimple. Now any finite-dimensional simple Lie algebra has trivial Frattini subalgebra (in characteristic zero), and this property carries over to direct sums. □

We can even recover Marshall's result that $F(L) \triangleleft L$ if $\dim L$ is finite and the field is algebraically closed of characteristic zero. By 5.2 it is sufficient to consider the semisimple case. But over an algebraically closed field of characteristic zero every semisimple Lie algebra is generated by elements x with x^* nilpotent. Then $F(L)$ is obviously invariant under all $\exp(x^*)$, hence an ideal of L. If the field is not algebraically closed this observation no longer has effect, for there may not be any non-zero x with x^* nilpotent.

Theorem 5.4 If L is ideally finite over a field of characteristic zero, then $F(L)$ is locally nilpotent.

Proof: $L/\nu(L)$ is the direct sum of $\sigma(L)/\nu(L)$ and a semisimple Levi subalgebra Λ, by corollary 4.4. It is obvious from this that $F(L/\nu(L)) = 0$, hence that $F(L) \leq \nu(L)$ and is locally nilpotent. □

Ideally finite Lie algebras have 'good Frattini structure' in a sense very close to that of [65] or [1] chapter 12. To state this precisely we define two more subalgebras. We say that H/K is a <u>chief factor</u> of L if $H, K \triangleleft L$ and there is no ideal of L lying strictly between H and K. We define

$$\psi(L) = \bigcap C_L(H/K)$$

where the intersection is over all chief factors H/K. We also define $\tilde{\nu}(L)$ by

$$\tilde{\nu}(L)/F(L) = \nu(L/F(L))$$

(noting that $F(L) \triangleleft L$ when L is ideally finite).

<u>Theorem 5.5</u> <u>If L is ideally finite over a field of characteristic zero, then</u>

$$\nu(L) = \psi(L) = \tilde{\nu}(L).$$

<u>Proof</u>: Pick any chief series for L, that is, a series (L_α) such that the factors $L_{\alpha+1}/L_\alpha$ are chief factors. Then $\psi(L)$ centralizes each factor of this series, and it follows that $\psi(L)$ is locally nilpotent, and hence easily that $\psi(L) \leq \nu(L)$. Now let U/V be a chief factor of L. Since L is ideally finite U/V is finite-dimensional. Let $N = \nu(L)$. If

$$N+V/V \cap U/V = 0$$

then $[U,N] \leq V$. If the intersection is not 0 then $U/V \leq N+V/V$. Since U/V has finite dimension we have $U/V \leq \zeta_n(N+V/V)$ for some integer n ([63] p.319 lemma 3.3.4) so U/V is central in N+V/V, and again $[U,N] \leq V$. Hence $N \leq \psi(L)$, and so we have $\nu(L) = \varphi(L)$.

It remains to prove that $\tilde{\nu}(L) = \nu(L)$, which we do as follows. We claim that $H \leq L$ is locally nilpotent if and only if $H+K/K$ is nilpotent for all ideals $K \triangleleft L$ of finite codimension. This is an easy consequence of lemma 3.3. Now apply this with $H = \tilde{\nu}(L)$. Then $H/F(L) = \nu(L/F(L))$. For each K of finite codimension we have that

$$(H+F(L)+K)/(F(L)+K)$$

is nilpotent. But $F(L/K) = F(L)+K/K$, so it follows that $H+K/K$ is nilpotent modulo $F(L/K)$. By Barnes and Newell [7] theorem 1.2 we have $H+K/K$ nilpotent. Hence H is locally nilpotent, and so $\tilde{\nu}(L) \leq \nu(L)$. The reverse inclusion is obvious. □

We shall make no use of the Frattini subalgebra in the remainder of these notes, but in view of its appearance in the finite-dimensional theory and its connections with formations, it deserves a mention here.

6 Levi subalgebras

A <u>Levi subalgebra</u> of an ideally finite Lie algebra is a subalgebra which complements the radical. Thus Λ is a Levi subalgebra of L if (writing $\dot{+}$ for a split extension)

$$L = \sigma(L) \dot{+} \Lambda.$$

It follows that Λ is isomorphic to $L/\sigma(L)$ so is semisimple. The following characterization of Levi subalgebras implies their existence as a corollary.

<u>Theorem 6.1</u> <u>Let L be ideally finite over k. Then Λ is a Levi subalgebra of L if and only if Λ is a maximal semisimple subalgebra.</u>

<u>Proof</u>: Clearly every Levi subalgebra is maximal semisimple. Let M be a maximal semisimple subalgebra of L, and suppose for a contradiction that $L \neq \sigma(L) + M$. Then there exists a finite-dimensional ideal F of L such that $F \not\leq \sigma(L)+M$. Let $C = C_L(F)$, which is an ideal of finite codimension in L. Then $M \cap C$ is of finite codimension in M, so we can write

$$M = (M \cap C) \oplus M'$$

where M' is a finite-dimensional ideal of M (by lemma 4.9). Now $M \cap C$ centralizes F and M', hence centralizes F+M'. Since $F \not\leq \sigma(L)+M$ we have $F \neq \sigma(F)+M'$, so M' is not a Levi subalgebra of F+M'. We may find a Levi subalgebra M" of F+M' which contains M' (since F+M' is finite-dimensional). Now $M \cap C$ centralizes M", so

$$M_1 = (M \cap C) + M''$$

is semisimple and contains M properly. This contradicts the maximality of M. Hence $L = \sigma(L)+M$. But obviously $\sigma(L) \cap M = 0$, so M is a Levi subalgebra. □

<u>Corollary 6.2</u> <u>Every semisimple subalgebra of</u> L <u>is contained in a Levi subalgebra</u>.

<u>Proof</u>: Use the obvious Zorn's lemma argument. □

Levi subalgebras are well-behaved with regard to ideals and homomorphisms:

<u>Proposition 6.3</u> <u>Let</u> L <u>be ideally finite over</u> k, <u>with a Levi subalgebra</u> Λ. <u>Let</u> I <u>be an ideal of</u> L. <u>Then</u>

(a) $\Lambda \cap I$ <u>is a Levi subalgebra of</u> I,

(b) $(\Lambda+I)/I$ <u>is a Levi subalgebra of</u> L/I.

<u>Proof</u>: (a) We may without loss of generality quotient out $\sigma(I)$ and assume I is semisimple, in which case we have to show that $I \leq \Lambda$. But $I+\Lambda$ is semisimple, for if S is a locally soluble ideal of $I+\Lambda$ then $\frac{S+I}{I}$ is a locally soluble ideal of the semisimple algebra $\frac{\Lambda+I}{I}$, so $S \leq I$; whence $S = 0$ since I is semisimple. Now theorem 6.1 implies $I \leq \Lambda$ as required.

The proof of (b) is obvious. □

In such an argument it is often useful to quotient by $\sigma(I)$, for the following reason:

<u>Proposition 6.4</u> <u>Every semisimple ideal of an ideally finite Lie algebra over</u> k <u>is a direct summand</u>.

Proof: Let I be a semisimple ideal of the ideally finite algebra L. There exists a Levi subalgebra Λ of L containing I. Then $\Lambda = I \oplus I'$ for some I', so $L = \sigma(L)+\Lambda = (\sigma(L)+I')+I$, and

$$[\sigma(L)+I', I] \leq \sigma(L) \cap I = 0. \qquad \square$$

It is clear that in fact $\sigma(L)+I' = C_L(I)$.

The methods used so far in the last few chapters have been very elementary 'local nonsense' or Zorn's lemma arguments. In order to obtain the more interesting conjugacy theorems it will be necessary to introduce new, less elementary, techniques. To motivate these we briefly consider the problem of a conjugacy theorem for Levi subalgebras. Let Λ_1 and Λ_2 be Levi subalgebras of the ideally finite Lie algebra L. It is not true that for some $\alpha=\exp(x^*)$ we have $\Lambda_1^{\alpha} = \Lambda_2$, as is easily seen by considering a direct sum of infinitely many copies of a Lie algebra having two distinct Levi subalgebras. The situation is highly analogous to that for Sylow subgroups of FC-groups (Kuroš [45] p.169) and it is natural to attempt to use similar methods. If $\{F_i\}_{i \in I}$ is the set of all finite-dimensional ideals of L we find that $\Lambda_{1i} = \Lambda_1 \cap F_i$ and $\Lambda_{2i} = \Lambda_2 \cap F_i$ are Levi subalgebras of F_i. By the finite-dimensional theory we have an automorphism α_i of F_i such that $\Lambda_{1i}^{\alpha_i} = \Lambda_{2i}$. The problem is to fit all of these α_i together to obtain an automorphism α of L. We can find suitable α_i which fix all $F_j \leq F_i$. If we let $\mathfrak{A}_i$ be the set of all α_i fixing those $F_j \leq F_i$ and mapping Λ_{1i} to Λ_{2i} we can use restriction to define a projective limit system $\{\mathfrak{A}_i, f_{ij}\}$. By analogy with Kuroš, we wish to show that proj lim $\mathfrak{A}_i \neq \emptyset$, because we can make any $\alpha \in$ proj lim $\mathfrak{A}_i$ act as an automorphism of L with the property $\Lambda_1^{\alpha} = \Lambda_2$.

In the case studied by Kuroš, the sets $\mathfrak{A}_i$ are finite and this is well-known to imply proj lim $\mathfrak{A}_i \neq \emptyset$. In our case $\mathfrak{A}_i$ is not in general finite. However it does have a natural structure as an algebraic variety which, it turns out, enables the argument to be pushed through to the desired conclusion. The next chapter will set up the machinery required.

7 Projective limits of varieties

We begin by recalling the pertinent facts about projective (or inverse) limits. Let I be a directed set with partial ordering $\leq$, so that if $i, j \in I$ then there exists $k \in I$ such that $i \leq k$ and $j \leq k$. A <u>projective limit system</u> (of topological spaces) over I is a set of topological spaces $\{X_i\}_{i \in I}$ together with continuous maps $f_{ij}: X_i \to X_j$ ($i, j \in I$; $i \geq j$) satisfying a compatibility condition

$$f_{ij} f_{jk} = f_{ik}$$

whenever $i \geq j \geq k$; $i, j, k \in I$. In other words, the diagrams

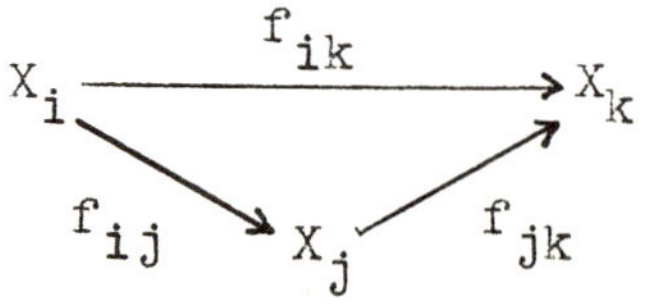

commute whenever they exist. We further require that each f_{ii} is the identity map on X_i for each $i \in I$. We denote such a system by

$$\underset{\sim}{X} = \{X_i, f_{ij}\}.$$

The <u>projective limit</u>

$$X = \text{proj lim } X_i = \text{proj lim } \underset{\sim}{X}$$

is the subset of $\prod_{i \in I} X_i$ consisting of those elements (x_i) such that $f_{ij}(x_i) = x_j$ for all i, j I with $i \geq j$. We can make X into a topological space by taking the subspace topology induced by the Tychonoff topology on $\prod_{i \in I} X_i$. Equivalently we may take the weakest topology in which all the restrictions of coordinate projections

$$f_i : X \to X_i$$

are continuous (see Bourbaki [10] p.52 §4 n°.4).

As we remarked above, Kuroš's 'projection set' method is essentially the fact that a projective limit of non-empty finite sets is non-empty. This, as is well known, is a special case of the theorem that a projective limit of non-empty compact Hausdorff topological spaces is non-empty. Now any algebraic variety is compact in the Zariski topology, but not Hausdorff. It does, however, satisfy a separation axiom weaker than the Hausdorff axiom, in that it is T_1. In this situation we may apply a result given in Bourbaki [11] p.138. In the statement of the theorem, we say that a map $f:X \to Y$ of topological spaces X,Y is <u>closed</u> if $f(F)$ is closed in Y for every closed set $F \subseteq X$.

<u>Theorem 7.1</u> <u>Let $\{X_i, f_{ij}\}$ be a projective limit system of topological spaces X_i such that</u>

(i) <u>Each X_i is non-empty, compact, and T_1,</u>

(ii) <u>The maps f_{ij} are closed.</u>

<u>Then $X = \text{proj lim } X_i$ is not empty.</u>

<u>Proof</u>: We use Zorn's lemma. Let $\mathcal{S}$ be the set of subsystems $\{A_i\}$ of X such that each A_i is a non-empty closed subset of X_i and $f_{ij}(A_i) \subseteq A_j$ for all $i,j \in I$ with $i \geq j$. Define a partial ordering $\ll$ on $\mathcal{S}$ such that $\{A_i\} \ll \{B_i\}$ if $A_i \subseteq B_i$ for all $i \in I$. The compactness of the X_i implies that $\mathcal{S}$ satisfies the hypotheses of Zorn's lemma with respect to the inverse ordering $\gg$, so there exists $\{A_i\} \in \mathcal{S}$, minimal with respect to $\ll$.

By (ii) it follows that

$$B_i = \bigcap_{j \geq i} f_{ji}(A_j)$$

is closed in A_i. It follows that $\{B_i\} \in \mathcal{S}$. The minimality of $\{A_i\}$ implies that $B_i = A_i$ for all $i \in I$, or equivalently that the maps f_{ij}, when restricted to A_i, are surjective.

For a fixed $i \in I$ choose $x_i \in A_i$. If we let

$$C_j = f_{ji}^{-1}(x_i) \cap A_j \qquad (j \geq i)$$

$$C_j = A_j \qquad (j \not\geq i)$$

then the T_1 property implies that $\{C_i\} \in \mathcal{S}$. By minimality $A_i = C_i = \{x_i\}$. Thus for all $i \in I$ we have $A_i = \{x_i\}$, so that $(x_i) \in \prod_{i \in I} X_i$ belongs to proj lim X_i, and the theorem is proved. □

In fact, X = proj lim X_i is compact (see [57] theorem 2.1) but we do not need this fact here. The results of Bourbaki [11] p.138 are still more general than theorem 7.1.

In order to apply this theorem to the algebraic varieties which arise in conjugacy problems, we have to circumvent a final obstacle. We have noted in chapter 2 that, although morphisms between varieties are continuous relative to the Zariski topology, they are not necessarily closed. However, theorem 7.1 requires closed maps. Theorem 2.4 rescues us, at the price of changing the topology, as follows.

Let G be a (linear) algebraic group over k. Let $\mathcal{Z}$ denote the Zariski topology on G. Define a topology $\mathcal{W}$ as follows: a closed subbase consists of all cosets xH of $\mathcal{Z}$-closed subgroups H of G. Thus a closed set in $\mathcal{W}$ is a finite union

$$x_1H_1 \cup \ldots \cup x_nH_n$$

of cosets, where $x_1, \ldots, x_n \in G$ and $H_1, \ldots, H_n$ are algebraic subgroups of G.

Each coset $x_i H_i$ is $\mathcal{Z}$-closed in G, so the topology $\mathcal{W}$ is weaker than the Zariski topology $\mathcal{Z}$. Just how much weaker can be illustrated by letting G be the additive group of k^2, in which case $\mathcal{W}$ has as subbase all points, affine lines, and the plane k^2 itself; whereas $\mathcal{Z}$ has all algebraic curves as well.

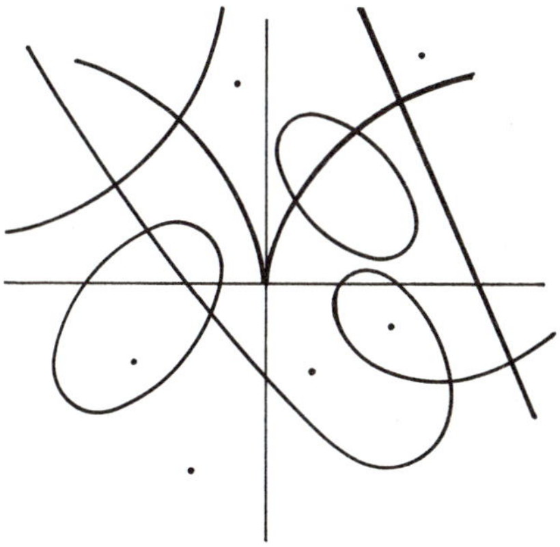

The Zariski topology

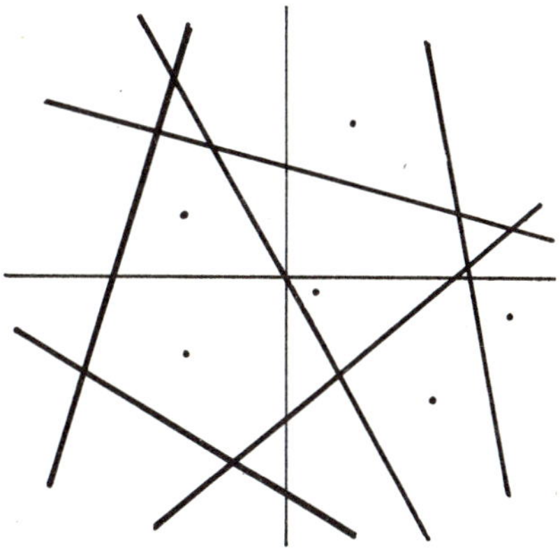

The $\mathcal{W}$-topology

It follows at once that $\mathcal{W}$ is compact and T_1 (the identity subgroup is algebraic, and its cosets are points). Of course, similar results hold for the topologies $\mathcal{W}'$ obtained by taking cosets Hx instead of xH, or $\mathcal{W}''$ generated by $\mathcal{W}$ and $\mathcal{W}'$. The reason for considering $\mathcal{W}$ is:

Lemma 7.2 Let G and K be algebraic groups over k, and let $\alpha: G \to K$ be an algebraic group morphism. If G and K are equipped with the $\mathcal{W}$-topology, then α is continuous and closed.

<u>Proof</u>: We know that α is $\mathcal{Z}$-continuous. If H is a $\mathcal{Z}$-closed subgroup of K then $\alpha^{-1}(H)$ is a $\mathcal{Z}$-closed subgroup of G. Let $x \in K$, and pick $z \in \alpha^{-1}(x)$ if this is non-empty. Clearly

$$\alpha^{-1}(xH) = z.\alpha^{-1}(H)$$

which is $\mathcal{W}$-closed in G. Passing to unions, we see that α is $\mathcal{W}$-continuous.

Now let L be a $\mathcal{Z}$-closed subgroup of G, and $g \in G$. Then $\alpha(gL) = \alpha(g)\alpha(L)$. By theorem 2.4, $\alpha(L)$ is $\mathcal{Z}$-closed in K, hence $\alpha(g)\alpha(L)$ is $\mathcal{W}$-closed. Again we may pass to finite unions, so α is $\mathcal{W}$-closed. □

The example given just after theorem 2.4 shows that it is essential here to have k algebraically closed. However, we have not used the fact that k has characteristic zero.

We can generalize further, to a <u>homogeneous space</u> G/H where G is an algebraic group and H an algebraic subgroup of G. As a set, this consists of all cosets xH ($x \in G$). It can be given a varietal structure (not affine!) but for our purposes a topological structure suffices. There is a natural map

$$\nu : G \to G/H$$
$$x \mapsto xH.$$

This defines a quotient topology on G/H relative to the $\mathcal{W}$-topology on G, which we shall call the $\mathcal{W}$-topology on G/H. If $\alpha : G \to K$ is an algebraic group morphism, and if H and L are algebraic subgroups of G,K respectively, such that $\alpha(H) \subseteq L$, then there is an induced map

$$\tilde{\alpha}: G/H \to K/L.$$

with $\mathcal{W}$-topologies on G/H and K/L, $\tilde{\alpha}$ is also closed and continuous. We define a <u>coset variety</u> over k to be any $\mathcal{W}$-closed subset of a homogeneous space G/H. A map $\tilde{\alpha}: G/H \to K/L$, or its restriction to a coset variety contained in G/H, will be called <u>affine</u>, if it is induced by an algebraic group morphism $\alpha: G \to K$ such that $\alpha(H) \subseteq L$. It is now easy to prove:

<u>Lemma 7.3</u> <u>Affine maps between coset varieties are continuous and closed relative to the $\mathcal{W}$-topology.</u> □

Combining this with theorem 7.1 we obtain a variant of a theorem of Serre [53] p.15 (see also Bourbaki [11]):

<u>Theorem 7.4</u> <u>Let $\{X_i, f_{ij}\}$ be a projective limit system, where the X_i are coset varieties over k and the maps f_{ij} are affine. Suppose the X_i are non-empty. Then proj lim X_i is non-empty.</u>

□

In fact we apply this theorem only to coset varieties which are themselves cosets xH, or homogeneous spaces G/H. The above formulation combines the two types of application.

8 Conjugacy of Levi and Borel subalgebras

In this chapter we shall apply projective limits to prove conjugacy theorems for Levi and Borel subalgebras of ideally finite Lie algebras. More refined versions of the conjugacy theorems will be proved in later chapters; here we try to keep the method as simple as possible.

If L is a Lie algebra and G is a group of automorphisms of L, we say that subsets X and Y of L are G-conjugate if there exists $\alpha \in G$ such that $Y = X^{\alpha}$.

Let L be ideally finite over k, with $\{F_i\}_{i \in I}$ its set of finite-dimensional ideals, partially ordered by inclusion. For each i we let $\mathcal{G}_i$ be the group of all automorphisms of F_i which fix setwise all $F_j \leq F_i$. If $j \leq i$ then restriction defines a map

$$f_{ij}: \mathcal{G}_i \to \mathcal{G}_j.$$

Since

$$\mathcal{G}_i = \bigcap_{j \leq i} N_{\mathrm{Aut}(F_i)}(F_j)$$

it follows that $\mathcal{G}_i$ is an algebraic group. We equip it with the $\mathcal{W}$-topology. The maps f_{ij} are algebraic group morphisms, and

$$\{\mathcal{G}_i, f_{ij}\}$$

is a projective limit system. We define

$$\mathcal{G}(L) = \text{proj lim } \mathcal{G}_i.$$

This is a group (coordinatewise operations) and acts naturally as a group of automorphisms of L. If $\alpha = (\alpha_i) \in \mathcal{G}(L)$, and $x \in F_i$, we can define $x^{\alpha} = x^{\alpha_i}$. The compatibility condition

$f_{ij}(\alpha_i) = \alpha_j$ ensures that this is well-defined, and it is easy to verify that the map $x \mapsto x^{\alpha}$ is an automorphism of L. We identify $\mathcal{G}(L)$ with its image in Aut(L). We will call any group which is a projective limit of algebraic groups a proalgebraic group (a term used in slightly different senses by other authors). Every proalgebraic group carries a natural compact topology, namely the $\mathcal{W}$-topology.

Theorem 8.1 Let L be ideally finite over k. Then any two Levi subalgebras of L are $\mathcal{G}(L)$-conjugate.

Proof: Let Λ_1 and Λ_2 be Levi subalgebras of L, and define $\{F_i\}_{i \in I}$ as above. Then for each $i \in I$, $\Lambda_{1i} = \Lambda_1 \cap F_i$ and $\Lambda_{2i} = \Lambda_2 \cap F_i$ are Levi subalgebras of F_i (6.3(a)). Define

$$\mathcal{B}_i = \{\alpha \in \mathcal{G}_i : \Lambda_{1i}{}^{\alpha} = \Lambda_{2i}\}.$$

The theorem of Mal'cev - Harish-Chandra (Jacobson [39] p.92) shows that $\mathcal{B}_i \neq \emptyset$. If $\alpha \in \mathcal{B}_i$ it is clear that

$$\mathcal{B}_i = \alpha . N_{\mathcal{G}_i}(\Lambda_{2i})$$

and hence is a coset variety. With f_{ij} defined by restriction we have $f_{ij}(\mathcal{B}_i) \subseteq \mathcal{B}_j$ whenever $j \leq i$. It follows that $\{\mathcal{B}_i, f_{ij}\}$ is a projective limit system of non-empty coset varieties, and the maps f_{ij} are clearly affine; so by theorem 7.4

$$\mathcal{B} = \text{proj lim } \mathcal{B}_i \neq \emptyset.$$

If $\beta \in \mathcal{B}$ then it follows that $\Lambda_{1i}{}^{\beta} = \Lambda_{2i}$ for all $i \in I$, and hence that $\Lambda_1{}^{\beta} = \Lambda_2$. $\square$

The diagram below illustrates the argument.

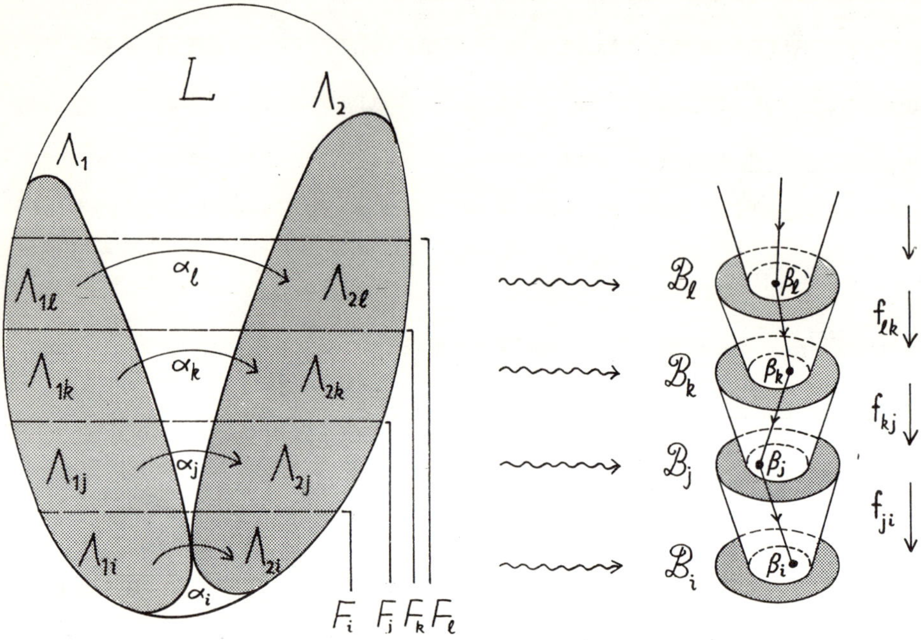

A *Borel subalgebra* of an ideally finite Lie algebra is a maximal locally soluble subalgebra. It is clear from Zorn's lemma that Borel subalgebras exist. They may be characterized in terms of the Borel subalgebras of finite-dimensional simple Lie algebras.

Proposition 8.2 *Let* L *be ideally finite over* $\mathfrak{k}$, *and let* $L/\sigma(L)$ *be decomposed as* $\bigoplus_{d\in D} S_d$ *where each* S_d *is finite-dimensional and simple. Then the Borel subalgebras of* L *are precisely the complete inverse images in* L *of subalgebras* $\bigoplus_{d\in D} B_d$ *where* B_d *is a Borel subalgebra of* S_d *for each* $d \in D$.

Proof: It is easy to see that subalgebras of this type are always Borel subalgebras. Conversely, suppose that B is a Borel subalgebra of L. Then $B+\sigma(L)$ is locally soluble, so $\sigma(L) \leq B$. Hence $B/\sigma(L)$ is a Borel subalgebra of $L/\sigma(L)$. By considering coordinate projections onto the S_d it follows that B is contained in a subalgebra of the type described in the theorem, hence by maximality equal to that subalgebra. □

Corollary 8.3 *Let L be ideally finite over* $\mathfrak{k}$, *and let* Λ *be a Levi subalgebra of L with decomposition* $\Lambda = \bigoplus_{d\in D} S_d$ *into simple subalgebras. As the* B_d *range independently over the Borel subalgebras of* S_d $(d \in D)$, *the algebras*

$$\sigma(L) + \bigoplus_{d\in D} B_d$$

are precisely the Borel subalgebras of L. □

A conjugacy theorem for Borel subalgebras will follow exactly as for Levi subalgebras, once we obtain suitable behaviour on intersecting with ideals.

Lemma 8.4 *Let L be ideally finite over* $\mathfrak{k}$, *and let K be an ideal of L. If B is a Borel subalgebra of L then* $B \cap K$ *is a Borel subalgebra of K.*

Proof: Let Λ be a Levi subalgebra of L, with the usual decomposition $\Lambda = \bigoplus_{d\in D} S_d$. Then $B = \sigma(L) + \bigoplus_{d\in D} B_d$ for Borel subalgebras $B_d \leq S_d$. Now $K = \sigma(K) + (\Lambda \cap K)$ by 6.3(a), and since $\Lambda \cap K \triangleleft \Lambda$ we have $\Lambda \cap K = \bigoplus_{d\in E} S_d$ for some subset $E \subseteq D$. Now $B \cap K = \sigma(K) + \bigoplus_{d\in E} B_d$ which is a Borel subalgebra of K.

□

Theorem 8.5 *Let L be ideally finite over k. Then any two Borel subalgebras of L are $\mathcal{G}(L)$-conjugate.*

Proof: Imitate that of theorem 8.1, using lemma 8.4 in place of lemma 6.3(a). The $\mathcal{B}_i$ are non-empty since obviously elements of $\mathcal{E}(F_i)$ fix setwise all ideals of F_i, and Borel subalgebras of F_i are $\mathcal{E}(F_i)$-conjugate by 1.15. □

There is also a homomorphism-invariance property of Borel subalgebras, which will be needed later:

Lemma 8.6 *Let L be ideally finite over k, and let K be an ideal of L. If B is a Borel subalgebra of L then $(B+K)/K$ is a Borel subalgebra of L/K.*

Proof: We have $\sigma(K) = \sigma(L) \cap K$ (by 4.6) which is an ideal of L. Since $\sigma(L) \leq B$ we may pass to the quotient $L/\sigma(K)$, or equivalently assume K semisimple. By proposition 6.4 K is a direct summand of L, complemented by $\sigma(L)+\Lambda$ where Λ is semisimple. Then $\Lambda+K$ is a Levi subalgebra of L, and using corollary 8.3 we obtain the desired result. □

9 Locally inner automorphisms

In this section we introduce the group of locally inner automorphisms $\mathcal{L}(L)$ of an ideally finite Lie algebra L, which plays the role of $\mathcal{E}(L)$ in finite dimensions. To motivate the construction of this group consider a finite-dimensional Lie algebra L having distinct Cartan subalgebras C, C'. For an index set I (infinite) take isomorphic copies L_i of L, and let C_i, C'_i be the images of C ,C'. Now

$$D = \bigoplus_{i \in I} L_i$$

has Cartan subalgebras

$$E = \bigoplus_{i \in I} C_i,$$

$$E' = \bigoplus_{i \in I} C'_i.$$

No product of automorphisms exp(x*) for elements $x \in D$ with x* nilpotent can map E to E', because such automorphisms fix all but a finite number of summands L_i. On the other hand, if $\alpha \in \mathcal{E}(L)$ is such that $C^{\alpha} = C'$, then we can define an automorphism β of D to have the effect of α on each L_i, and then $E^{\beta} = E'$. Now β is <u>locally</u> like an automorphism in $\mathcal{E}$, for when restricted to any subalgebra of the form

$$L_{i_1} \oplus \dots \oplus L_{i_n} = X \text{ (say)}$$

with n finite, β equals $(\alpha,\dots,\alpha)$ and lies in $\mathcal{E}(X)$.

This fact, and the analogy with FC-groups,prompts the following definition. An automorphism α of L is <u>locally inner</u> if, given any finite set of elements $x_1,\dots,x_n \in L$ we can find a finite-dimensional subalgebra X of L containing $x_1,\dots,x_n$ and an automorphism $\beta \in \mathcal{E}(X)$ such that $x_i^{\alpha} = x_i^{\beta}$

for $i = 1,\ldots,n$.

In this section we shall develop methods for constructing locally inner automorphisms, and obtain some basic properties of $\mathcal{L}$.

Let L be ideally finite over $\mathfrak{k}$, having a finite residual system $\{K_j\}_{j\in J}$, and order J by reverse inclusion. Whenever $i \leq j$ there is a natural homomorphism

$$\pi_{ji}: L/K_j \to L/K_i.$$

For each $j \in J$ we let $\mathcal{C}_j$ be the set of all automorphisms of L/K_j leaving invariant all K_i/K_j for $i \leq j$. It is clear that $\mathcal{C}_j$ is an algebraic group, and that if $i \leq j$ then π_{ji} induces a morphism

$$p_{ji}: \mathcal{C}_j \to \mathcal{C}_i.$$

Clearly we have a projective limit system $\{\mathcal{C}_j, p_{ji}\}$, and we let $\mathcal{C}(L)$ be its limit.

Next, let $\mathcal{D}_j$ be the set of all $\beta \in \mathcal{C}_j$ such that β leaves invariant I/K_j for every $I \triangleleft L$ with $I \geq K_j$. This is an algebraic subgroup of $\mathcal{C}_j$, and $\{\mathcal{D}_j, p_{ji}|_{\mathcal{D}_j}\}$ is also a projective limit system. We let $\mathcal{D}(L)$ be its limit.

We will show that the elements of $\mathcal{D}(L)$ act naturally as automorphisms of L, and in fact that there is a natural monomorphism $\mathcal{D}(L) \to \mathcal{G}(L)$, where $\mathcal{G}(L)$ is defined as in section 8. For let $\gamma \in \mathcal{D}(L)$, so that $\gamma = (\gamma_j)_{j\in J}$ where each γ_j is an automorphism of L/K_j fixing all ideals of L/K_j, and where if $i \leq j$ then

$$p_{ji}(\gamma_j) = \gamma_i. \qquad (*)$$

Let $x_1,\ldots,x_n \in L$. We can find a finite-dimensional ideal X of L containing $x_1,\ldots,x_n$, and by lemma 3.4 we have $X \cap K_k = 0$ for some $k \in J$. The natural injection $e: X \to L/K_k$ can now be used to pull back γ_k, giving an automorphism of X: we use this to define the action of γ on $x_1,\ldots,x_n$. By (*) the action is well-defined, and we obtain the desired monomorphism.

By refining the construction slightly we can obtain a locally inner automorphism. Let

$$\mathcal{H}_j = \mathcal{E}(L/K_j) \subseteq \mathcal{D}_j.$$

Now the lifting property of $\mathcal{E}$ (specifically surjectivity of the induced map) shows that

$$\{\mathcal{H}_j, p_{ji}|_{\mathcal{H}_j}\}$$

is a projective limit system, so that

$$\mathcal{H}(L) = \text{proj lim}\, \mathcal{H}_j \subseteq \mathcal{D}(L).$$

It follows easily that under the monomorphism $\mathcal{D}(L) \to \mathcal{G}(L)$ the elements of $\mathcal{H}(L)$ act as locally inner automorphisms.

The set of all locally inner automorphisms of L is a group, and we denote it by

$$\mathcal{L}(L).$$

Thus we have a canonical monomorphism $\mathcal{H}(L) \to \mathcal{L}(L)$.

To obtain functorial properties of $\mathcal{L}$ similar to those of $\mathcal{E}$ we require an alternative description. Let $\{F_i\}_{i \in I}$ be the set of all finite-dimensional ideals of L and order I by inclusion. For each $i \in I$ and each $j \geq i$ the group $\mathcal{E}(F_j)$ induces by restriction automorphisms of F_i. Let $\mathcal{L}_{ji}$ be the resulting subgroup of $\mathrm{Aut}(F_i)$, and denote the restriction map by

$$r_{ji}: \mathcal{E}(F_j) \to \mathcal{L}_{ji}.$$

The extension property for $\mathcal{E}$ shows that $\mathcal{E}(F_i) \subseteq \mathcal{L}_{ji}$ for all $j \geq i$. We define

$$\mathcal{L}_i = \bigcup_{j \geq i} \mathcal{L}_{ji}.$$

The extension property implies that the set of all $\mathcal{L}_{ji}$, for $j \geq i$, is directed by inclusion, so that $\mathcal{L}_i$ is a subgroup of $\mathrm{Aut}(F_i)$ fixing setwise all smaller F_k. But $\mathcal{E}(F_j)$ is a connected algebraic group (corollary 2.12) and it follows that each $\mathcal{L}_{ji}$ is a connected algebraic subgroup of $\mathrm{Aut}(F_i)$. Now theorem 2.9, which says that $\mathrm{Aut}(F_i)$ has the ascending chain condition for connected algebraic subgroups, implies that $\mathcal{L}_i = \mathcal{L}_{i_o i}$ for some $i_o \geq i$, and hence that $\mathcal{L}_i$ is a connected algebraic group.

If f_{ji} is the map induced by restriction $F_j \to F_i$ then we clearly have

$$f_{ji}(\mathcal{L}_j) = \mathcal{L}_i$$

for all $j \geq i$. Therefore we obtain a projective limit system $\{\mathcal{L}_i, f_{ji}\}$. Obviously

$$\mathcal{L}(L) = \text{proj lim } \mathcal{L}_i.$$

We may now prove theorems analogous to some of Stonehewer [69] giving lifting and extension properties for $\mathcal{L}$.

<u>Theorem 9.1</u> <u>Let</u> L <u>be ideally finite over</u> k, <u>with</u> $H \leq L$.

(a) <u>Every element</u> $\sigma' \in \mathcal{L}(H)$ <u>extends to an element</u> $\sigma \in \mathcal{L}(L)$.

(b) <u>Every epimorphism</u> $L \to L'$ <u>induces an epimorphism</u> $\mathcal{L}(L) \to \mathcal{L}(L')$.

<u>Proof</u>: Let $\{F_i\}_{i \in I}$ be the set of all finite-dimensional ideals of L, ordering I by inclusion.

(a) For each $i \in I$ define

$$\mathcal{X}_i = \{\alpha \in \mathcal{L}_i : \alpha|_{H \cap F_i} = \sigma'|_{H \cap F_i}\},$$

and note that $H \cap F_i$ is σ'-invariant. Obviously $\mathcal{X}_i$ is a coset variety. To see that $\mathcal{X}_i \neq \emptyset$ choose $x_1, \dots, x_n$ spanning $H \cap F_i$. We can find $j \in I$ such that $x_1, \dots, x_n \in H \cap F_j$, and there exists $\tau \in \mathcal{E}(H \cap F_j)$ such that

$$x_1^{\sigma'} = x_1^{\tau}, \dots, x_n^{\sigma'} = x_n^{\tau}.$$

If we extend τ to τ' on F_j and put $\alpha = \tau'|_{F_i}$ then $\alpha \in \mathcal{X}_i$.

Letting f_{ji} denote restriction, we obtain a projective limit system $\{\mathcal{X}_i, f_{ji}\}$. If we pick $\sigma \in \text{proj lim } \mathcal{X}_i$ then $\sigma|_H = \sigma'$, and $\sigma \in \mathcal{L}(L)$.

(b) The proof is similar, only now we let $\mathcal{X}_i$ be the set of all $\alpha \in \mathcal{L}_i$ such that α induces the same automorphism as σ' on L', and choose $\sigma \in \text{proj lim } \mathcal{X}_i$. It is obvious that the induced map is surjective. □

10 Existence and conjugacy of Cartan subalgebras

Having (at least for the moment) disposed of Borel and Levi subalgebras, it is natural to consider Cartan subalgebras. The problems here are more delicate. Even the 'correct' definition is at the outset not too obvious. Having settled on a definition, the proof of existence is non-trivial. And the conjugacy requires different methods, because the intersection of a Cartan subalgebra with an ideal is not in general a Cartan subalgebra of the ideal. We can, however, obtain a grip on Cartan subalgebra through the finite-dimensional quotients of a given Lie algebra, because Cartan subalgebras have 'functorial' behaviour with respect to homomorphisms; and the machinery of the previous section is designed to take advantage of this fact. It therefore seems reasonable to define Cartan subalgebras in such a way that their homomorphism-invariance is immediately apparent. The most satisfactory way to do this is via formation theory.

Thus let L be ideally finite over k. We say that a subalgebra C of L is a <u>Cartan subalgebra</u> of L if

(i) C is locally nilpotent,

(ii) Whenever $C \leq H \leq L$, $K \triangleleft H$, and H/K is locally nilpotent, then K+C = H.

In formation-theoretic terms, C is a <u>locally nilpotent projector</u> of L. The spirit of the definition is that if C is a Cartan subalgebra of L and $C \leq H \leq L$ then C is a Cartan subalgebra of H, <u>and</u> that if K is an ideal then C+K/K is a Cartan subalgebra of L/K. Part (ii) of the definition combines the two forms of behaviour.

We shall show later that Cartan subalgebras are precisely the locally nilpotent self-idealizing subalgebras. Although this seems more familiar as a definition, it is technically harder to handle.

The definition has a number of useful consequences.

<u>Lemma 10.1</u> *Let* L *be ideally finite over* k, *with* $X \triangleleft L$, *and let* C *be a Cartan subalgebra of* L. *Then*

(a) C *is a maximal locally nilpotent subalgebra of* L.

(b) $C+X/X$ *is a Cartan subalgebra of* L/X.

(c) *If* $C \leq H \leq L$ *then* C *is a Cartan subalgebra of* H.

(d) *If* C'/X *is a Cartan subalgebra of* L/X *and* C'' *is a Cartan subalgebra of* C', *then* C'' *is a Cartan subalgebra of* L.

(e) C *contains the centre of* L.

(f) *If* C' *contains the centre* Z *of* L *and if* C'/Z *is a Cartan subalgebra of* L/Z, *then* C' *is a Cartan subalgebra of* L.

(g) *If* $C \leq H \leq L$ *then* H *is self-idealizing in* L.

<u>Proof</u>: (a) Let H be locally nilpotent, $C \leq H$. Then $H/0$ is locally nilpotent, so $H = 0+C = C$.

(b) Suppose $C+X/X \leq H/X \leq L/X$ where $K/X \triangleleft H/X$ is such that $(H/X)/(K/X)$ is locally nilpotent. Then H/K is locally nilpotent, $C \leq H \leq L$, so $H = C+K$. Therefore $H/X = C+X/X + K/X$ as required.

(c) If $C \leq H' \leq H$ and H'/K is locally nilpotent then, since $H' \leq L$, we have $H' = C+K$ as required.

(d) Suppose $C'' \leq H \leq L$, $K \triangleleft H$, and H/K is locally nilpotent. Then $H+X/K+X \cong H/(H \cap X)+K$ is locally nilpotent, and so

$H+X = C'+K$. Hence

$$H = K + (C' \cap H). \qquad (*)$$

Also $(C' \cap H)/(C' \cap K) \cong (C' \cap H)+K/K = H/K$ by (*), so is locally nilpotent. Since C'' is a Cartan subalgebra of C' we have $C' \cap H = (C' \cap K)+C''$, and it follows from (*) that $H = K+C''$.

(e) If Z is the centre, then $C+Z$ is locally nilpotent and part (a) applies.

(f) This follows from (d).

(g) If $x \in H \setminus I_L(H)$ then $\langle x\rangle+H/H$ is locally nilpotent, so $\langle x\rangle+H = H+C = H$, hence we have a contradiction. □

The argument of Tomkinson [71] yields the following lemma:

<u>Lemma 10.2</u> <u>Let</u> L <u>be a residually finite ideally finite Lie algebra over</u> $\mathfrak{k}$, <u>and let</u> $\{K_j\}_{j\in J}$ <u>be the set of ideals of</u> L <u>of finite codimension. For each</u> $j \in J$ <u>there exists a subalgebra</u> C_j <u>of</u> L <u>such that</u>

(a) $K_j \leq C_j$,

(b) <u>If</u> $K_i \geq K_j$ <u>then</u> $C_i \geq C_j$,

(c) C_j/K_j <u>is a Cartan subalgebra of</u> L/K_j.

<u>Proof</u>: Let $\mathcal{C}_j$ be the set of Cartan subalgebras of L/K_j. Order J by reverse inclusion as usual. If $j \leq i$ the natural homomorphism $L/K_i \to L/K_j$ induces a map

$$p_{ij}: \mathcal{C}_i \to \mathcal{C}_j$$

because of the homomorphism-invariance of Cartan subalgebras in finite dimensions. We shall give the $\mathcal{C}_j$ the structure of coset varieties, in such a way that the p_{ij} become affine.

Let C be a Cartan subalgebra of a finite-dimensional Lie algebra F over $\mathfrak{k}$. Since Aut(F) is transitive on the set of Cartan subalgebras of F, this set is in bijective correspondence with the homogeneous space

$$Aut(F)/N_{Aut(F)}(C).$$

This is a coset variety, so we can pull back its structure to define a coset variety structure on the set of Cartan subalgebras of F. To show that the p_{ij} are affine it is sufficient to show that the $\mathcal{W}$-topology induced on the set of Cartan subalgebras of F is independent of the choice of C, for we may then choose C's related by a homomorphism. A Cartan subalgebra of F will be of the form C^{α} for some $\alpha \in Aut(F)$. Since

$$N_{Aut(F)}(C^{\alpha}) = (N_{Aut(F)}(C))^{\alpha^{-1}},$$

and since inner automorphisms are $\mathcal{Z}$-homeomorphisms of Aut(F) to itself, it follows that the $\mathcal{W}$-topology is indeed independent of C.

Now $\{\mathcal{C}_j, p_{ij}\}$ is a projective limit system of non-empty coset varieties with affine maps, so we may find

$$(C_j) \in \text{proj lim } \mathcal{C}_j.$$

These C_j satisfy (a), (b), and (c).

The existence of Cartan subalgebras is a consequence of the following result:

<u>Theorem 10.3</u> <u>With the hypotheses and notation of lemma 2.2, the subalgebra</u> $C = \bigcap_{j \in J} C_j$ <u>is a Cartan subalgebra of</u> L.

Proof: If $x \in A$, where A is a finite-dimensional Lie algebra, we shall write

$$A_0(x) = \{a \in A: ax^{*n} = 0 \text{ for some } n\}.$$

This is the null-component of A regarded as $\langle x\rangle$-module under adjoint action. If B is a Cartan subalgebra of A then 1.10 implies that $B = A_0(x)$ for a regular element $x \in A$. If $D \triangleleft A$, $y \in A$, and ϕ is any homomorphism of A, then it is easy to verify that $D_0(y)\phi = (D\phi)_0(y\phi)$. (A more general result is proved in section 12.) In particular if $X \triangleleft A$ then $B+X/X = (A/X)_0(x+X)$.

Let F be an arbitrary finite-dimensional ideal of L. Choose $i \in J$ such that $F \cap K_i = 0$, and let $C_F = C_i \cap F$. Let x_i be any element of L for which $C_i/K_i = (L/K_i)_0(x_i+K_i)$. For sufficiently large n we have

$$(C_i \cap F)x_i^{*n} \subseteq K_i \cap F = 0,$$

so that $C_i \cap F \leq F_0(x_i)$. But $F_0(x_i)+K_i/K_i \leq (L/K_i)_0(x_i+K_i) = C_i/K_i$, so $F_0(x_i) \leq C_i \cap F$, and we get

$$C_F = C_i \cap F = F_0(x_i),$$

this being independent of the choice of x_i.

Let $j \geq i$ and let x_j+K_j be an element of L/K_j such that

$$(L/K_j)_0(x_j+K_j) = C_j/K_j.$$

Applying the natural homomorphism $L/K_j \to L/K_i$ we find that

$$C_i/K_i = (L/K_i)_0(x_j+K_i),$$

and

$$C_i \cap F = F_0(x_j) = C_j \cap F.$$

Thus C_F is independent of the choice of i. It follows that if $F_1 \leq F_2$ are finite-dimensional ideals of L, then $C_{F_1} \leq C_{F_2}$.

Hence for $C = \bigcap_{j \in J} C_j$ we have $C = \bigcup_F C_F$, so C is a locally nilpotent subalgebra of L.

Suppose now that $C \leq H \triangleright K$, with H/K locally nilpotent. Let F be any finite-dimensional ideal of L, and choose $i \in J$ such that $F \cap K_i = 0$. Since $C+K_i = C_i$, there exists $x_i \in C$ such that x_i+K_i belongs to L/K_i and has C_i/K_i as its null-component. Then x_i^* induces a nilpotent map on H/K, and hence a nilpotent map on $H \cap F/K \cap F$. Thus

$$H \cap F = (K \cap F)+(H \cap F)_0(x_i) \leq (K \cap F)+C_F,$$

and equality is clear. Since L is the union of all such F, it follows that H = K+C. This proves that C is a Cartan subalgebra of L. □

<u>Corollary 10.4</u> <u>Let</u> L <u>be ideally finite over</u> $\mathfrak{k}$. <u>Then</u> L <u>has a Cartan subalgebra.</u>

<u>Proof</u>: Let Z be the centre of L. Then L/Z is residually finite, so by 10.2 and 10.3 it has a Cartan subalgebra C/Z. Then C is a Cartan subalgebra of L by 10.1(f). □

The next corollary is useful in proving that a given subalgebra is a Cartan subalgebra.

<u>Corollary 10.5</u> <u>Let</u> L <u>be ideally finite over</u> $\mathfrak{k}$, <u>and let</u> C <u>be a subalgebra of</u> L <u>such that</u> C+K/K <u>is a Cartan subalgebra of</u> L/K <u>for every ideal</u> K <u>of finite codimension in</u> L. <u>If</u> Z <u>is the centre of</u> L, <u>then</u> C+Z <u>is a Cartan subalgebra of</u> L.

<u>Proof</u>: This is immediate from 10.3 and 10.1(f). □

With the machinery available, a conjugacy theorem is now straightforward.

<u>Theorem 10.6</u> <u>Let</u> L <u>be ideally finite over</u> $\mathfrak{k}$. <u>Then any two Cartan subalgebras of</u> L <u>are</u> $\mathcal{L}$(L)-<u>conjugate</u>.

<u>Proof</u>: Let Z be the centre of L. It is contained in both Cartan subalgebras. Since elements of $\mathcal{L}(L/Z)$ lift to $\mathcal{L}(L)$ we may work modulo Z, and hence assume L is residually finite. Suppose that in this situation the two Cartan subalgebras are C_1 and C_2. Then for any ideal K_j of L of finite codimension it follows that C_1+K_j/K_j and C_2+K_j/K_j are Cartan subalgebras of L/K_j. Define

$$\mathfrak{X}_j = \{\alpha \in \mathcal{E}(L/K_j) : (C_1+K_j/K_j)^{\alpha} = C_2+K_j/K_j\}$$

which is a non-empty coset variety. If $\beta \in$ proj lim $\mathfrak{X}_j$ then β is locally inner, and

$$(C_1+K_j)^{\beta} = C_2+K_j \qquad (j \in J).$$

Hence

$$(\bigcap_{j\in J} C_1+K_j)^{\beta} = \bigcap_{j\in J} C_2+K_j.$$

The right-hand side contains C_2. Considering its images modulo each K_j it is easily seen to be locally nilpotent, so by maximality equal to C_2. Similarly the left-hand side is C_1^{β}. □

11 Improved conjugacy theorems

The main aim in this section is to bring theorems 8.1 and 8.5 into line with theorem 10.6, replacing the group $\mathcal{G}(L)$ by $\mathcal{L}(L)$. For Borel subalgebras, the method of theorem 10.6 apply without further ado, to yield:

Theorem 11.1 If L is ideally finite over $\mathfrak{k}$ then any two Borel subalgebras of L are $\mathcal{L}(L)$-conjugate. □

This result may be strengthened, as in Winter [78] p.99. We define a Borel-Cartan pair of L to be a pair (B,C) where B is a Borel subalgebra of L and C is a Cartan subalgebra of B. First we need:

Lemma 11.2 If L is ideally finite over $\mathfrak{k}$ and (B,C) is a Borel-Cartan pair of L, then C is a Cartan subalgebra of L.

Proof: Suppose to begin with that L is semisimple, so that $L = \bigoplus_{m \in M} S_m$ for finite-dimensional simple ideals S_m. We know that $B = \bigoplus_{m \in M} B_m$ for Borel subalgebras B_m of S_m. The projection C_m of C on B_m is a Cartan subalgebra of B_m, and since C is maximal locally nilpotent, $C = \bigoplus_{m \in M} C_m$. The lemma is true in finite dimensions, so C_m is a Cartan subalgebra of S_m for each $m \in M$. To prove C is a Cartan subalgebra of L, suppose $C \leq H \triangleright K$ where H/K is locally nilpotent. For each finite-dimensional ideal of L of the form

$$S = S_{m_1} \oplus \dots \oplus S_{m_t}$$

we know that $C \cap S$ is a Cartan subalgebra of S, so we have

$$H \cap S = (K \cap S) + (C \cap S).$$

It follows that $H = K+C$, so C is a Cartan subalgebra of L.

In the general case, let $R = \sigma(L)$. Then $B \geq R$ and B/R is a Borel subalgebra of L/R. Now $C+R/R$ is a Cartan subalgebra of B/R by 10.1(b). Since C is a Cartan subalgebra of $C+R \leq B$ (lemma 10.1(c)) it follows that C is a Cartan subalgebra of L by 10.1(d). □

<u>Theorem 11.3</u> <u>If</u> L <u>is ideally finite over</u> $\mathfrak{k}$, <u>then any two Borel-Cartan pairs of</u> L <u>are</u> $\mathcal{L}(L)$-<u>conjugate.</u>

<u>Proof</u>: Let (B,C) and (B',C') be Borel-Cartan pairs of L. By theorem 11.1 there exists $\alpha \in \mathcal{L}(L)$ such that $B^{\alpha} = B'$. Now C^{α} and C' are Cartan subalgebras of B', so by theorem 10.6 there exists $\gamma \in \mathcal{L}(B')$ such that $C^{\alpha\gamma} = C'$. By theorem 9.1(a) we can extend γ to $\beta \in \mathcal{L}(L)$, and then $B'^{\beta} = B'$ so we have $B^{\alpha\beta} = B'$, $C^{\alpha\beta} = C'$. □

To treat Levi subalgebras similarly, the essential point is to obtain $\mathcal{E}(L)$-conjugacy when L is finite-dimensional. This may be deduced from the standard theorems as follows:

<u>Lemma 11.4</u> <u>Let</u> L <u>be a finite-dimensional Lie algebra over</u> $\mathfrak{k}$. <u>Then any two Levi subalgebras of</u> L <u>are</u> $\mathcal{E}(L)$-<u>conjugate.</u>

<u>Proof</u>: By induction on the dimension, using the lifting and extension properties of $\mathcal{E}$, we may assume that $L = A \dotplus \Lambda$ where A is an abelian ideal, Λ is semisimple, and A is irreducible as Λ-module. The theorem of Mal'cev and Harish-Chandra

(Jacobson [39] p.92) implies that all Levi subalgebras of L are conjugate by automorphisms of the form exp(a*) where $a \in A$.

Let B be the additive subgroup of A generated by all strongly nilpotent elements of L which lie inside A. Then B is a vector space invariant under all automorphisms of L. It follows that B is an ideal of L (by theorem 1.3, or by looking at the elements $x \in A$ with x* nilpotent, and using exponentials, together with the fact that for such elements x, x* is nilpotent on L (Humphreys [38] p.30)). Irreducibility of A implies that B = 0 or B = A. If B = 0 then any $y \in L$ for which y* is semisimple (diagonalizable) must centralize A; so A, being generated by such y, centralizes A. This means that A is the unique Levi subalgebra of L, and the result is trivial. Otherwise B = A, so $a \in A$ can be written as

$$a = a_1 + \dots + a_n$$

where each a_i is strongly nilpotent. But now

$$\exp(a^*) = \prod_{i=1}^{n} \exp(a_i^*)$$

since A is abelian, and the right-hand side lies in $\mathcal{E}(L)$. □

With this established, previous methods easily lead to

<u>Theorem 11.5</u> *Let L be ideally finite over $\mathfrak{k}$. Then any two Levi subalgebras of L are $\mathcal{L}(L)$-conjugate.* □

12 Chevalley-Jordan and Fitting decompositions

Let V be a vector space over k, and $f: V \to V$ a linear map. We shall say that f is <u>pure</u> (semisimple, diagonalizable) if V has a basis of f-eigenvectors (or equivalently if V is spanned by f-eigenvectors), and f is <u>nil</u> if every $v \in V$ is annihilated by some power of f (which may depend on v). If there exists a polynomial $q(t) \in k[t]$, $q(t) \neq 0$, such that $q(f) = 0$, then f is <u>algebraic</u>. There then exists a unique monic q of smallest degree such that $q(f) = 0$, the <u>minimum polynomial</u> of f. We say that f is <u>cleft</u> if we can write

$$f = f_p + f_n \tag{1}$$

where f_p is pure, f_n is nil, and $f_p f_n = f_n f_p$.

Lie algebras of algebraic linear transformations are studied by Curtis [15]. Algebraic transformations inherit a useful property from finite dimensions:

<u>Lemma 12.1</u> <u>If $f: V \to V$ is algebraic, then f is cleft. The transformations f_p and f_n in (1) are unique. There exist polynomials g, h $\in k[t]$ with zero constant term, such that $f_p = g(f)$, $f_n = h(f)$. Thus f_p and f_n leave invariant any subspace of V which f leaves invariant, annihilate any subspace which f annihilates, and commute with any linear transformation of V with which f commutes.</u>

<u>Proof</u>: Let $\lambda_1, \ldots, \lambda_k$ be the distinct zeros of the minimum polynomial

$$q(t) = (t-\lambda_1)^{m_1} \ldots (t-\lambda_k)^{m_k}$$

of f. Let $V_i = \ker (f-\lambda_i)^{m_i}$. We claim that V is the direct sum of the V_i, that each V_i is f-invariant, and that the minimum polynomial of $f|_{V_i}$ is $(t-\lambda_i)^{m_i}$.

For each $i = 1,\ldots,k$ let

$$q_i(t) = q(t)/(t-\lambda_i)^{m_i}.$$

Then the q_i are coprime, and we may find polynomials $p_1,\ldots,p_k$ such that

$$p_1 q_1+\ldots+p_k q_k = 1.$$

If $v \in V$ it follows that

$$v = vp_1(f)q_1(f)+\ldots+vp_k(f)q_k(f).$$

Now

$$vp_i(f)q_i(f)(f-\lambda_i)^{m_i} = vp_i(f)q(f) = 0$$

so that $vp_i(f)q_i(f) \in V_i$. It follows that V is the sum of the V_i. The other assertions are easy.

Apply the Chinese remainder theorem to find, in $k[t]$, a polynomial g such that

$$g(t) \equiv \begin{cases} \lambda_i & \text{mod } (t-\lambda_i)^{m_i} \quad (1\le i\le k) \\ 0 & \text{mod } t \,. \end{cases}$$

(The last congruence is superfluous if some $\lambda_i = 0$; if not the relevant set of polynomials remains coprime.)

Define

$$f_p = g(f)$$
$$f_n = f - g(f) = h(f)$$

where $h(t) = t - g(t)$. Both g and h have zero constant term. Now f_p and f_n leave invariant each V_i. On a given V_i, f_p is scalar multiplication by λ_i, so f_p is pure. It is clear that $f_n^{m_i} = 0$ when restricted to V_i (it acts in 'upper triangular' fashion), so f_n is nil(potent). And f_p and f_n obviously commute.

The remainder of the theorem follows immediately, except for uniqueness. To prove this, let $f = f_p + f_n$ satisfy the relevant conditions. Let $\{\lambda_i\}$ be the set of eigenvalues of f_p, with corresponding eigenspaces E_i. Then

$$E_i(f_p-\lambda_i) = 0.$$

Let $e \in E_i$, and choose e so that $ef_n^{\ m} = 0$. Then

$$e(f-\lambda_i)^m = e((f_p-\lambda_i)+f_n)^m = 0$$

by the binomial theorem. Hence e lies in the subspace V_i defined as above and λ_i is the corresponding eigenvalue. It follows that f_p acts as scalar multiplication by λ_i on V_i, so is as constructed; hence f_n is also as constructed and both are unique. □

For algebraic f we call the unique transformations f_p, f_n with the given properties the <u>pure part</u> and <u>nilpotent part</u> of f. Note that an algebraic nil transformation is nilpotent; in particular a nil transformation of finite rank is nilpotent. The decomposition $f = f_p+f_n$ is the <u>Chevalley-Jordan decomposition</u>, or <u>cleaving</u>, of f.

Mal'cev [46] introduced a kind of cleaving within a Lie algebra (of finite dimension) which is very useful for obtaining structural information. We can extend Mal'cev's ideas to ideally finite Lie algebras as follows. If L is a Lie algebra over k and $x \in L$ we say that x is <u>ad-algebraic</u>, <u>ad-pure</u>, or <u>ad-nil(potent)</u> according as the adjoint map x* is algebraic, pure, or nil(potent) on L. If there exist x_p, x_n L such that $x_p + x_n = x$, $[x_p, x_n] = 0$, x_p^* is pure and x_n^* is nil, then we say that x is <u>ad-cleft</u> in L. It follows that $x^* = x_p^* + x_n^*$ is the cleaving of x^*. Hence lemma 12.1 implies that for ad-algebraic x the components x_p and x_n are unique <u>modulo the centre of</u> L, since the maps x_p^*, x_n^* are unique. (This slight lack of uniqueness will be the cause of minor technical excursions later, explaining the occasional lack of directness in proofs.) If every $x \in L$ is ad-cleft we say that L is <u>cleft</u>.

There is a generalization which we must take into account. If M is an L-module we say that L is M-<u>cleft</u> if every $x \in L$ can be written as $x = x_p + x_n$ $(x_p, x_n \in L)$ with $[x_p, x_n] = 0$, in such a way that the maps induced on M by multiplication by x_p, x_n are pure, nil respectively. Thus L is cleft if and only if it is L-cleft, the action being the adjoint action. It is necessary <u>ab initio</u> to distinguish different types of cleaving, although it will eventually turn out that many such distinctions are without effect.

Examples in finite dimensions clearly illustrate these ideas. Every nilpotent Lie algebra is cleft (trivially), as is every semisimple Lie algebra (a theorem, see Humphreys [38] p.24). The algebra $\langle a, b : [a, b] = a \rangle$ is cleft. For if $x = \lambda a + \mu b$ then either $\mu = 0$, when x^* is nilpotent; or $\mu \neq 0$ in which case x^* is pure with eigenvalues λ, 0 on the linearly independent eigenvectors a, x.

On the other hand,

$$L = \langle x,y,t: [x,y] = 0,\ [x,t] = x+y, [y,t] = y\rangle$$

is not cleft. With respect to the basis $\{x,y,t\}$ the matrix of t^* is

$$\begin{pmatrix} 1 & 1 & 0 \\ 0 & 1 & 0 \\ 0 & 0 & 0 \end{pmatrix}$$

so $(t^*)_p$ has matrix

$$\begin{pmatrix} 1 & 0 & 0 \\ 0 & 1 & 0 \\ 0 & 0 & 0 \end{pmatrix}.$$

However, no inner derivation $(\alpha x+\beta y+\gamma t)^*$ has this matrix, since

$$x(\alpha x+\beta y+\gamma t)^* = \gamma x+\gamma y \neq x$$

whatever the value of γ.

However, we can 'adjoin' to L elements which will act like $(t^*)_p$ and $(t^*)_n = t^* - (t^*)_p$, say by passing to the algebra

$$L' = \langle x,y,t,t_p: [x,y] = 0,\ [x,t] = x+y,\ [y,t] = y,$$
$$[x,t_p] = x,\ [y,t_p] = y,\ [t,t_p] = 0\rangle.$$

This algebra _is_ cleft.

This exemplifies a general phenomenon, noted by Mal'cev in finite dimensions. A non-cleft algebra can be extended to a cleft algebra, in which it is very tightly embedded. One of our main aims in the next few sections will be to extend Mal'cev's results to ideally finite Lie algebras. In the course of this we shall make precise the 'tight-fitting' quality of the embedding.

We turn our attention to a related idea. Let L be a Lie algebra over $\mathfrak{k}$, and let $\lambda: L \to \mathfrak{k}$ be a 1-dimensional representation of L. As in chapter 1 we may define <u>weight spaces</u> M_λ in an L-module M by:

$$M_\lambda = \{m \in M: \text{for all } x \in L \text{ there exists } n > 0 \text{ such that } m(x^* - \lambda(x))^n = 0\}$$

where now x^* denotes the map $M \to M$ for which $mx^* = mx$. We can extend theorem 1.8 to ideally finite Lie algebras if we impose a condition on M. Namely, say M is <u>locally finite</u> if every finite subset of M is contained in a finite-dimensional L-submodule. (The most important example is an ideally finite Lie algebra under adjoint action.) We have:

<u>Lemma 12.2</u> <u>Let L be a locally nilpotent Lie algebra over $\mathfrak{k}$ and let M be a locally finite L-module. Then M is the direct sum of its weight-spaces, and these are all L-submodules.</u>

<u>Proof</u>: Local nonsense. Every finite-dimensional submodule X of M is a module for $L/C_L(X)$, which is finite-dimensional nilpotent. Hence X is a direct sum of weight spaces for the $L/C_L(X)$-action. The L-action factors through this, so X is a direct sum of weight spaces for L. Therefore L is the sum of weight spaces M_λ. That this sum is direct can be proved either by adapting the usual argument in finite dimensions or by looking at a local system in M. □

Local nilpotence of L can obviously be weakened: all that is really needed is that every finite-dimensional quotient of L should be nilpotent.

The weight spaces are well behaved functorially: if N is a submodule of M then for each λ

$$N_\lambda = N \cap M_\lambda \qquad (2)$$

$$(M/N)_\lambda = (M_\lambda + N)/N. \qquad (3)$$

If L is ideally finite and H is a subalgebra which is locally nilpotent then L, with adjoint action, is a locally finite H-module and so decomposes into weight spaces L_λ. As in finite dimensions, weight-spaces multiply according to

$$[L_\lambda, L_\mu] \subseteq L_{\lambda+\mu}. \qquad (4)$$

This follows from the identity

$$[y,z](x^*-\alpha-\beta)^n = \Sigma_{i=0}^{n} \quad [y(x^*-\alpha)^{n-i}, z(x^*-\beta)^i]$$

which is easily proved by induction (cf. Jacobson [39] p.64).

The expression of the module M as a direct sum of weight spaces we shall call the Fitting decomposition

$$M = \bigoplus_\lambda M_\lambda$$

of M. If $m \in M$ spans a 1-dimensional L-submodule then we have

$$mx = \lambda(x)m$$

for all $x \in L$, where $\lambda: L \to \mathfrak{k}$ is a 1-dimensional representation. We shall say that m is an L-eigenvector with eigenvalue λ.

Lemma 12.3 Let $f: V \to V$ be a linear map such that V is a locally finite $\langle f \rangle$-module and f is pure. Then every weight space V_λ consists entirely of f-eigenvectors with eigenvalue λ.

Proof: Let $x \in V_\lambda$. Since f is pure, x is a sum of f-eigenvectors. Each f-eigenvector lies in a weight space, and the sum of the weight spaces is direct; hence x is an f-eigenvector with eigenvalue λ.

Corollary 12.4 *Let* $f:V \to V$ *be a linear map such that* V *is a locally finite* <f>*-module, and let* W *be a submodule*.

(a) *If* f *is pure then it induces pure maps on* W *and* V/W.

(b) *If* f *is nil(potent) then it induces nil(potent) maps on* V/W.

(c) *A cleaving of* f *on* V *induces a cleaving of the maps induced on* W *and* V/W.

Proof: Part (a) follows from 12.3 and equations (2) and (3). Parts (b) and (c) are obvious. □

13 Toral structure and Cartan subalgebras

A torus in a Lie algebra L is a subalgebra T such that every element of T is ad-pure in its action on L. An elementary but important fact is:

Lemma 13.1 Every torus of a locally finite Lie algebra over $\mathfrak{k}$ is abelian.

Proof: Let L be locally finite, having a torus T. We use the argument of Humphreys [38] p.35 to show that $t^*_T = 0$ for all $t \in T$. Now corollary 12.4(a) tells us that t^*_T is pure (local finiteness of L implies local finiteness as <t>-module) so it suffices to show that t^*_T has no non-zero eigenvalues. Suppose, for a contradiction, that $ut^*_T = \lambda u$ where $0 \neq u \in T$ and $0 \neq \lambda \in \mathfrak{k}$. Now $t = t_1 + \ldots + t_n$ where the t_i are linearly independent elements of T such that $[t_i, u] = \lambda_i t_i$ for $\lambda_i \in \mathfrak{k}$. We have

$$0 = -\lambda[u,u] = [t,u,u] = \Sigma \lambda_i^2 t_i$$

and linear independence implies $\lambda_i = 0$ for all i, so that $[t,u] = 0$ which contradicts $\lambda \neq 0$. □

A maximal torus in a Lie algebra L is a torus which is not properly contained in another torus. Zorn's lemma shows that every Lie algebra possesses a maximal torus and every torus is contained in a maximal torus. It is immediate that every maximal torus contains the centre.

The importance of tori stems, in part, from the following theorem.

Theorem 13.2 Let L be a cleft ideally finite Lie algebra over $\mathfrak{k}$. If T is a maximal torus of L then $C_L(T)$ is a Cartan subalgebra of L.

Proof: Let $C = C_L(T)$. By lemma 12.1 C is L-cleft (for if c belongs to C then so do c_p and c_n by the polynomial property) and since T is abelian, $T \leq C$. Now T contains every $c \in C$ for which c^* is pure on L, because $[T,C] = 0$ and therefore $T + \langle c \rangle$ is a torus for any such c. If $c \in C$ then $c = c_p + c_n$ is a cleaving in L. By the previous remark $c_p{}^*$ is zero on C, so $c^*|_C = c_n{}^*|_C$ which is nilpotent. Engel's theorem, applied to a local system for C, shows that C is locally nilpotent.

Now suppose that $C \leq H \triangleright K$ where H/K is locally nilpotent. Then C, H, and K are all T-modules. If we decompose into weight spaces, the weight space corresponding to the weight 0 is C. If $K+C \neq H$ then H/K has a non-zero weight as T-module, so there exist $t \in T$, $x \in H$, $0 \neq \lambda \in \mathfrak{k}$ such that $[x,t] \equiv \lambda x$ (mod K). Then $xt^{*n} \equiv \lambda^n x \not\equiv 0$ (mod K) for all n, which contradicts H/K locally nilpotent.

Therefore C is a locally nilpotent projector, i.e. a Cartan subalgebra of L. □

As a corollary we can generalize a conjugacy theorem of Mal'cev [46]:

Theorem 13.3 Let L be cleft ideally finite over $\mathfrak{k}$. Then any two maximal tori of L are $\mathcal{L}(L)$-conjugate.

Proof: Let T and T' be the maximal tori. Their centralizers C and C' are Cartan subalgebras of L, so by 10.6 there exists $\alpha \in \mathcal{L}(L)$ such that $C' = C^{\alpha}$. The proof of theorem 13.2 shows that T is precisely the set of ad-pure elements of C and T' the set of ad-pure elements of C'. Since automorphisms preserve ad-purity, the result follows. □

We may now use an embedding process, very close to that used by Mal'cev in [46], to obtain an alternative proof for the existence of Cartan subalgebras in ideally finite Lie algebras, not relying on projective limit methods. Theorem 13.2 already does this in the cleft case. By lemma 11.2 (whose proof does not involve projective limits) we may further reduce the general case to the locally soluble case. The details required are also necessary for the construction of the 'cleft envelope' of an ideally finite Lie algebra, and will occupy the rest of this section.

Recall that for any vector space V the Lie algebra $\underset{\sim}{F}(V)$ consisting of all linear maps $V \to V$ of finite rank is locally finite (theorem 3.1). Further, it is cleft.

Lemma 13.4 The Lie algebra $\underset{\sim}{F}(V)$ over k is cleft for any vector space V.

Proof: It is obvious that $\underset{\sim}{F}(V)$ is V-cleft: the problem is to deduce that it is $\underset{\sim}{F}(V)$-cleft (with adjoint action). Now if $f \in \underset{\sim}{F}(V)$ then by lemma 12.1 there exists a unique V-cleaving $f = f_p + f_n$ where f_p and f_n are polynomials in f without constant term. It follows that $f_p, f_n \in \underset{\sim}{F}(V)$. We claim that

$$f^* = f_p^* + f_n^*$$

is an ad-cleaving in $\underset{\sim}{F}(V)$. Now $[f_p^*, f_n^*] = [f_p, f_n]^* = 0$, so all we need prove is that f_p^* is pure on $\underset{\sim}{F}(V)$ and f_n^* is nilpotent.

If $g, h \in \underset{\sim}{F}(V)$ then induction shows that

$$gh^{*m} = \Sigma_{i=0}^{m} \quad (-1)^i \binom{m}{i} h^i g h^{m-i}$$

so if $h^r = 0$ then $h^{*2r} = 0$. It follows that f_n^* is nilpotent.

If $g \in \underset{\sim}{F}(V)$ is pure on V we can choose a basis $\{v_i\}_{i \in I}$ of V consisting of g-eigenvectors, so that $v_i g = \lambda_i v_i$ for $\lambda_i \in \mathcal{k}$. The elementary transformations e_{ij} $(i, j \in I)$ defined by

$$v_k e_{ij} = \delta_{ki} v_j \qquad (k \in I)$$

(Kronecker delta) form a basis for $\underset{\sim}{F}(V)$. A simple computation (familiar to anyone who has found the roots of a simple Lie algebra of type A_ℓ) shows that

$$e_{ij} g^* = e_{ij} g - g e_{ij} = (\lambda_j - \lambda_i) e_{ij}$$

so the e_{ij} are g^*-eigenvectors and g^* is pure. Hence f_p^* is pure. □

<u>Corollary 13.5</u> <u>Any Lie subalgebra of $\underset{\sim}{F}(V)$ which contains the pure and nilpotent parts of each of its elements (thought of as transformations of V) is cleft.</u> □

The next result is proved exactly as in finite dimensions:

Lemma 13.6 *Let L be a Lie algebra over* $\mathfrak{k}$ *and let d be a derivation of L such that L is a locally finite* $\langle d\rangle$*-module. Then the pure and nil parts of d are derivations of L.*

Proof: Let d_p, d_n be the pure and nil parts of d. Write L as a direct sum of weight-spaces for d>:

$$L = \bigoplus_{\alpha} L_{\alpha}.$$

The identity which proves formula (4) of chapter 12 shows that $[L_{\alpha}, L_{\beta}] \subseteq L_{\alpha+\beta}$. Now d_p acts as scalar multiplication by $\alpha = \alpha(d)$ on each L_{α}. If we take $x \in L_{\alpha}$, $y \in L_{\beta}$, then $[x,y]$ belongs to $L_{\alpha+\beta}$ and so

$$[x,y]d_p = (\alpha+\beta)[x,y]$$

whereas

$$[xd_p, y] + [x, yd_p] = [\alpha x, y] + [x, \beta y] = (\alpha+\beta)[x,y]$$

which shows d_p is a derivation. Now $d_n = d - d_p$ is also a derivation. □

Next we construct an interesting algebra of derivations of an ideally finite Lie algebra L over $\mathfrak{k}$. Let $\{F_i\}_{i \in I}$ be the set of all finite-dimensional ideals of L. Let $\Delta(L)$ be the subalgebra of Der(L) consisting of the derivations which fix setwise every ideal of L, so that $\Delta(L) \geq \text{Inn}(L)$. For each $i \in I$ define D_i to be the set of all $d \in \Delta(L)$ such that

$$Ld \subseteq F_i \tag{1}$$

$$C_L(F_i)d = 0. \tag{2}$$

If $x \in F_i$ then clearly $x^* \in D_i$.

<u>Lemma 13.7</u> <u>With the above notation, every</u> D_i <u>is a finite-dimensional ideal of</u> $\Delta(L)$.

<u>Proof</u>: Since F_i has finite dimension, $C_L(F_i)$ has finite co-dimension, and we can find a finite-dimensional vector space complement W_i to $C_L(F_i)$ in L. Each $d \in D_i$ is uniquely determined by its restriction to W_i (by (2)), and condition (1) implies that

$$\dim D_i \leq \dim \operatorname{Hom}(W_i, F_i) < \infty.$$

Let $d \in D_i$, $x \in \Delta(L)$. We have

$$\text{(i)} \qquad L[d,x] \subseteq Ldx + Lxd \subseteq F_i x + Ld \subseteq F_i + F_i = F_i.$$

$$\text{(ii)} \qquad C_L(F_i)[d,x] \subseteq C_L(F_i)dx + C_L(F_i)xd \subseteq 0x + C_L(F_i)d \subseteq 0,$$

remembering that $C_L(F_i) \lhd L$ so is fixed by x.

Hence $D_i \lhd \Delta(L)$ for all $i \in I$. □

We now define an ideal $\Gamma(L)$ of $\Delta(L)$ by

$$\Gamma(L) = \Sigma_{i \in I} D_i.$$

If $i,j,k \in I$ and $F_k = F_i + F_j$ then it is clear that $D_i + D_j \leq D_k$, so that

$$\Gamma(L) = \bigcup_{i \in I} D_i. \qquad (3)$$

We can now show that L 'very nearly' embeds in a cleft ideally finite Lie algebra:

Theorem 13.8 Let L be ideally finite over k. Then $\Gamma(L)$ is a cleft ideally finite Lie algebra. The adjoint representation of L induces a homomorphism $*: L \to \Gamma(L)$ whose image is Inn(L) and kernel $\zeta_1(L)$.

Proof: By lemma 13.7, $\Gamma(L)$ is ideally finite. If d is an element of $\Gamma(L)$ then by (3) $d \in D_i$ for some i. By condition (1) d has finite rank, so $\Gamma(L) \leq F(L)$. By lemma 12.1 d has a cleaving $d = d_p + d_n$. By lemma 13.6, d_p and d_n are derivations of L, and lemma 12.1 implies they satisfy conditions (1) and (2) and therefore belong to D_i, hence to $\Gamma(L)$. Now corollary 13.5 implies that $\Gamma(L)$ is cleft.

The assertions about * are obvious. □

The point is that * embeds $L/\zeta_1(L)$ in a cleft ideally finite Lie algebra. For immediate purposes we can work modulo $\zeta_1(L)$ quite easily, and this embedding suffices. For later results we must face the problem of embedding L itself in a cleft ideally finite Lie algebra: we postpone discussion of this to chapter 14.

Let J be any ideal of L, let $j \in J$, and let $d \in \Delta(L)$. We have

$$[j^*, d] = (jd)^* \in J^*$$

where $J^* = \{x^* : x \in J\}$, so J^* is an ideal of $\Delta(L)$ and hence an ideal of $\Gamma(L)$.

The algebra $\Gamma(L)$ is a little too big for what we have in mind, and must be chopped down to size:

Lemma 13.9 *Let L be ideally finite over* $\mathfrak{k}$. *Then there exists a cleft subalgebra* $\hat{L}$ *of* $\Gamma(L)$ *such that* $L^* = \{x^* : x \in L\}$ *is contained in* $\hat{L}$, *and* $\hat{L}^2 = L^{*2}$.

If L is locally soluble, so is $\hat{L}$.

Proof: Let $\Gamma = \Gamma(L)$, which contains L^*. Define an increasing sequence of subalgebras

$$L^* = L_0 \leq L_1 \leq L_2 \leq \cdots$$

of Γ, as follows: each L_{i+1} is the subalgebra of Γ generated by all x_p and x_n for $x \in L_i$. Lemma 12.1 shows inductively that each L_i is an ideal of Γ. We put

$$\hat{L} = \bigcup_{i=0}^{\infty} L_i \triangleleft \Gamma.$$

We claim that

$$L_{i+1}^2 = L_i^2.$$

If $x \in L_i$ then since x_p^* is a polynomial in x^* with zero constant term,

$$[L_i, x_p] \subseteq \Sigma_{r=1}^{\infty} L_i x^{*r} \subseteq L_i^2,$$

and the same goes for x_n. Therefore $[L_{i+1}, L_i] \leq L_i^2$. Now a repetition of this argument shows that

$$L_{i+1}^2 = [L_{i+1}, L_{i+1}] \leq [L_{i+1}, L_i] \leq L_i^2.$$

It follows that $L_i^2 = L^{*2}$ for all i. But now

$$\hat{L}^2 = \bigcup_{i=0}^{\infty} L_i^2 = L^{*2}.$$

From the definition of the L_i, it follows that $\hat{L}$ is Γ-cleft, hence cleft.

If L is locally soluble so is $L^{*2} = \hat{L}^2$, and therefore $\hat{L}$ is locally soluble. □

We next give several characterizations of Cartan subalgebras, analogous to results in finite dimensions, which in particular substantiate our earlier assertion that the Cartan subalgebras of ideally finite Lie algebras are precisely the locally nilpotent self-idealizing subalgebras. It is convenient to introduce here the following concept. A subalgebra Q of a Lie algebra L is quasiabnormal if for every subalgebra U with $Q \leq U \leq L$ we have $U = I_L(U)$, that is, U is self-idealizing.

Theorem 13.10 Let L be ideally finite over k, and let C be a locally nilpotent subalgebra of L. Then the following are equivalent:

(a) C is a Cartan subalgebra of L.

(b) C is quasiabnormal in L.

(c) C is self-idealizing in L.

(d) C is equal to the 0-weight space of its adjoint representation on L.

Proof: We show that (a)$\Rightarrow$(b)$\Rightarrow$(c)$\Rightarrow$(d)$\Rightarrow$(a). Suppose (a) is true, and let $C \leq U \leq L$. Let $x \in I_L(U)$ and let $V = U+\langle x\rangle$. Then $U \triangleleft V$ and V/U is abelian, so the projector property tells us that $V = U+C = U$, since $C \leq U$. Hence $U = I_L(U)$, so (b) is true. Obviously (b) implies (c). To prove (c)$\Rightarrow$(d) assume $C = I_L(C) = L_0$, the 0-weight space. If $L_0 \neq C$ choose $x \in L_0$ such that $x \notin C$. Now x is contained in a finite-dimensional ideal X of L, and $X_0 = L_0 \cap X$ is the 0-weight space of X as C-module, or equivalently as a module for the finite-dimensional nilpotent algebra $L/C_L(X)$. Now $x \in X_o$, and it follows from theorem 1.8 that there exists an integer r such that $[x,_r C] = 0$. Choose m maximal subject to $[x,_m C] \not\subseteq C$. Then $[[x,_m C],C] \subseteq [x,_{m+1} C] \subseteq C$, so that $[x,_m C] \subseteq I_L(C) = C$. But this contradicts the definition of m. Therefore $C = L_o$ as required.

Finally we prove that (d) $\Rightarrow$ (a). This is proved in a fashion exactly analogous to the second half of theorem 13.2, looking at the C-module structure and a weight space decomposition, and will therefore be omitted. □

We shall also require a generalization of a theorem of Stitzinger [66]:

<u>Lemma 13.11</u> <u>Let</u> L <u>be locally soluble ideally finite over</u> k, <u>and let</u> C <u>be a subalgebra of</u> L. <u>Then the following are equivalent</u>:

(a) C <u>is a Cartan subalgebra of</u> L.

(b) C <u>is a maximal locally nilpotent subalgebra of</u> L <u>and</u> $L = \nu(L)+C$.

<u>Proof</u>: The projector property shows that (a) implies (b), since $L/\nu(L)$ is abelian (corollary 4.4). To obtain the converse we may appeal to theorem 13.10 and show that (b) implies that C is self-idealizing. Suppose for a contradiction that $I = I_L(C) > C$, and let $N = \nu(L)$. If $I \cap N = C \cap N$ then $I = I \cap (N+C) = (I \cap N)+C = (C \cap N)+C = C$, a contradiction. Therefore $I \cap N > C \cap N$. Pick $x \in (I \cap N) \setminus C$, and let $M = C+\langle x\rangle$.

Now x^* is nilpotent, and theorem 3.6 implies that $C = \zeta_\omega(C) = \bigcup_{r=1}^{\infty} \zeta_r(C)$. Hence if $x_1,\ldots,x_s \in C$ then $x_1,\ldots,x_s \in \zeta_m(C)$ for some m, and then $\zeta_m(C)+\langle x\rangle$ is nilpotent by a simple argument (cf. [64] lemma 2.1 p.15). Hence M is locally nilpotent, contrary to the maximality of C. Therefore $C = I_L(C)$ and so C is a Cartan subalgebra. □

We may now give a somewhat <u>ad hoc</u> construction for a Cartan subalgebra, not depending on projective limits:

Theorem 13.12 <u>Every ideally finite Lie algebra over $\mathfrak{k}$ has a Cartan subalgebra.</u>

<u>Proof</u>: Let L be ideally finite over $\mathfrak{k}$. Pick any Borel subalgebra B. The map $*: B \to \hat{B}$ (with $\hat{B}$ defined as in 13.9) embeds $B^* = B/\zeta_1(B)$ in a cleft locally soluble ideally finite algebra $\hat{B}$. Use theorem 13.2 to construct a Cartan subalgebra C of $\hat{B}$. Now $\hat{B}/\hat{B}^2$ is abelian, so the projector property implies that $\hat{B} = \hat{B}^2 + C$. Now 13.9 says that $\hat{B}^2 = B^{*2}$, so we have $\hat{B} = B^{*2} + C$. Intersecting with B^* and using the modular law, $B^* = B^{*2} + (B^* \cap C)$. Pulling back to B we find that $B = B^2 + D$ where D is locally nilpotent (since the kernel of the map $B \to B^*$ is central). Now $B^2 \leq \nu(B)$. If M is a maximal locally nilpotent subalgebra of L containing D (which exists by Zorn's lemma) then $B = \nu(B)+M$, and 13.11 implies that M is a Cartan subalgebra of B. Now theorem 11.2 shows that M is a Cartan subalgebra of L. □

14 An embedding theorem

In this chapter we shall embed any ideally finite Lie algebra in a cleft ideally finite Lie algebra in a canonical fashion. We begin with a technical result, weakening the conditions for cleftness. Say that a linear transformation f of a vector space V is <u>semicleft</u> if f = g+h where g, h are linear transformations of V, g is pure, and f is nil. This differs from a cleaving only in that g and h are not required to commute. Call a Lie algebra L <u>semicleft</u> if each $x \in L$ can be written x = p+n where $p, n \in L$, p^* is pure, and n^* is nil. A theorem of Mal'cev [46] p.233 shows that a soluble finite-dimensional semicleft Lie algebra over $\mathfrak{k}$ is cleft. We shall generalize this. First we need:

<u>Lemma 14.1</u> <u>Let L be locally soluble ideally finite over</u> $\mathfrak{k}$. <u>An element</u> $x \in L$ <u>is ad-nilpotent if and only if</u> $x \in \nu(L)$.

<u>Proof:</u> The assertion is true for L finite-dimensional (cf. Mal'cev [46] p.233) and follows for the ideally finite case by local nonsense. □

<u>Theorem 14.2</u> <u>A locally soluble ideally finite Lie algebra over</u> $\mathfrak{k}$ <u>is cleft if and only if it is semicleft.</u>

<u>Proof:</u> Let L be semicleft, locally soluble, and ideally finite over $\mathfrak{k}$. Put $Z = \zeta_1(L)$, $N = \nu(L)$. Since L/Z is residually finite it follows that N/Z is residually nilpotent, so that

$$\bigcap_{m=1}^{\infty} N^m \leq Z. \qquad (1)$$

Let $x \in L$, with

$$x = p+n \qquad (2)$$

where $p, n \in L$, p^* is pure, and n^* nilpotent. Let X be any finite-dimensional ideal of L containing p, n, and x. If we decompose X into weight spaces for $\langle p \rangle$ we obtain

$$n = n_o + n_1 + \dots + n_s \qquad (3)$$

where $n_i \in X$ for all $i = 0, \dots, s$, and $[n_i, p] = \lambda_i n_i$ for distinct eigenvalues $\lambda_i \in \mathfrak{k}$. We choose the notation to make $\lambda_o = 0$, possibly $n_o = 0$, but $n_1, \dots, n_s \neq 0$. If s were zero we would have $[n, p] = 0$ and (2) would be an ad-cleaving, so we may assume $s \neq 0$. Our aim will be to modify the semicleaving (2) until this situation obtains.

Since X has finite dimension the decreasing chain

$$X \cap N \geq X \cap N^2 \geq \dots \geq X \cap N^k \geq \dots$$

terminates, say at $X \cap N^t$. From (1) we have $X \cap N^t \leq Z$. Now lemma 14.1 implies that $n \in N$. Suppose in (3) we have

$$u = n_1 + \dots + n_s \quad N^k,$$

for an integer $k > 0$. The elements $n_1, \dots, n_s$ must lie in $X \cap N^k$, since this is a $\langle p \rangle$-submodule and so we may write u as a sum of p^*eigenvectors in $X \cap N^k$ and then plead uniqueness. Now $\lambda_1^{-1} n_1 + \dots + \lambda_s^{-1} n_s \in N^k$, so is ad-nilpotent, and we may form the automorphism

$$\alpha = \exp(\lambda_1^{-1} n_1 + \dots + \lambda_s^{-1} n_s)^*$$

of L. Now

$$x^{\alpha} = p^{\alpha} + n_o^{\alpha} + n_1^{\alpha} + \dots + n_s^{\alpha}$$

$$= p - \Sigma_{i=1}^{s} n_i + \Sigma_{j=0}^{s} n_j + \Sigma_{i,j} \lambda_i^{-1} [n_i, n_j] + \dots$$

$$= p + n_o + n'$$

where $n' \in X \cap N^{k+1}$ and $x^{\alpha} \in X$. Decompose n' into p^*-eigenvectors in $N^{k+1} \cap X$, to obtain $n' = n_o' + n_1' + \ldots + n_s'$ where the subscripts correspond to the same eigenvalues λ_i as before. Then each $n_i' \in N^{k+1} \cap X$, and

$$x^{\alpha} = p + (n_o + n_o') + (n_1' + \ldots + n_s')$$

where p^* is pure, $(n_o+n_o')^*$ is nilpotent and centralizes p, and $(n_1'+\ldots+n_s') \in X \cap N^{k+1}$.

Repeat this process until the superscript k has been raised to t. Since $X \cap N^t \leq Z$ we get

$$x^{\beta} = p + n_o'' + z$$

where β is an automorphism, and belongs to $\mathcal{E}(L)$, n_o'' is ad-nilpotent, $[p, n_o''] = 0$, and $z \in X \cap N^t \leq Z$. Then the decomposition

$$x = p^{\beta^{-1}} + (n_o''+z)^{\beta^{-1}}$$

is an ad-cleaving. Hence L is cleft. □

Note that the automorphism β belongs to the group $\mathcal{E}(L)$, which is defined for ideally finite Lie algebras in the obvious way, and is then a subgroup of $\mathcal{L}(L)$. Hence each semicleaving $x = p+n$ yields a cleaving $p^{\gamma}+n'$ for $\gamma \in \mathcal{E}(L)$. Note also that we have used only the fact that x is semicleft to deduce that x is cleft, rather than cleftness of every element of L.

Next we establish some easy properties of representations of semisimple algebras.

<u>Lemma 14.3</u> <u>Let</u> L <u>be semisimple ideally finite over</u> k , <u>and let</u> M <u>be a locally finite</u> L-<u>module</u>.

(a) <u>Every submodule of</u> M <u>is complemented</u>.

(b) M <u>is a direct sum of finite-dimensional irreducible</u> L-<u>modules</u>.

(c) <u>If</u> $x \in L$ <u>is ad-pure on</u> L <u>then it is ad-pure on</u> M, <u>and if</u> $x \in L$ <u>is ad-nil on</u> L <u>then it is ad-nil on</u> M.

(d) L <u>is</u> M-<u>cleft</u>.

<u>Proof</u>: Each finite-dimensional submodule F of M is a module for the finite-dimensional semisimple algebra $L/C_L(F)$, so is a direct sum of irreducible L-modules by theorem 1.12. Thus M is a sum of finite-dimensional irreducible submodules. Easy Zorn's lemma arguments prove (a) and (b). To prove (c) and (d) note that they are true in finite dimensions (Humphreys [38] p.29) and argue on a local system. □

We shall refer to (c) as the 'preservation of Chevalley-Jordan decomposition' for semisimple algebras.

In Mal'cev's work [46] on finite-dimensional Lie algebras he uses methods involving Lie groups which are not available to us in the present context. Instead we may substitute a generalization of a theorem of Barnes [5] p.278. This substitution also works in finite dimensions, affording an alternative argument to Mal'cev's.

Lemma 14.4 Let L be ideally finite over $\mathfrak{k}$, with $H \triangleleft L$, and let C be a Cartan subalgebra of H. Then

$$L = I_L(C) + H.$$

Proof: Let $I = I_L(C)$ and consider L/I as C-module. If $I+a \in L/I$ is annihilated by C then $[a,C] \subseteq I \cap H = I_H(C) = C$, so $a \in I$. Therefore C annihilates no non-zero element of L/I. Now L/I is a locally finite C-module. If its weight space for the weight 0 were non-zero, there would exist a non-zero C-eigenvector with eigenvalue 0, contrary to what we havs just proved. Hence L/I has trivial 0-weight space. On the other hand, C acts trivially on L/H. To avoid a contradiction in the action on L/(I+H) we must have I+H = L as required. □

This lemma, as Barnes [5] remarks, is analogous to the famous 'Frattini argument' in group theory and serves a similar purpose. We use it to prove:

Lemma 14.5 Let L be ideally finite over $\mathfrak{k}$, with radical S, and let Λ be a Levi subalgebra. Then Λ idealizes some Cartan subalgebra of S. If further S is cleft, Λ idealizes the maximal torus corresponding to that Cartan subalgebra.

Proof: Let C be any Cartan subalgebra of S. Then the above lemma implies that $L = S + I_L(C)$. If Λ_o is a Levi subalgebra of $I_L(C)$ then it must also be a Levi subalgebra of L, so L has a Levi subalgebra Λ_o idealizing C. Now there exists $\alpha \in \mathcal{L}(L)$ such that $\Lambda_o^{\alpha} = \Lambda$. Since α leaves S invariant, C^{α} is a Cartan subalgebra of S idealized by $\Lambda_o{}^{\alpha} = \Lambda$.

Now let S be cleft, so that $C^{\alpha} = C_S(T)$ for a maximal torus T. Then T is the unique maximal torus of S contained in C^{α} and consists of those $c \in C^{\alpha}$ for which $c^*{}_S$ is pure. Now Λ, being a direct sum of simple finite-dimensional algebras, is cleft, and the theorem on preservation of Chevalley-Jordan decomposition of semisimple algebras by representations applied to L (considered as adjoint Λ-module) shows that $x \in \Lambda$ is ad-nilpotent on Λ then it is ad-nilpotent on L. Now for each integer n, the automorphism $\exp(nx^*)$ of L leaves C^{α} invariant, since Λ idealizes C^{α}. Hence it also leaves T invariant. By proposition 1.2, x^* fixes T. Now the ad-nilpotent elements x generate Λ since Λ is semisimple, so Λ idealizes T as claimed. □

Now we may obtain a useful decomposition of a cleft locally soluble Lie algebra, which we call a <u>global semicleaving</u>:

<u>Lemma 14.6</u> <u>Let L be cleft locally soluble ideally finite over k. Then</u>

$$L = \nu(L) + T$$

<u>for any maximal torus T, and</u>

$$\nu(L) \cap T = \zeta_1(L).$$

<u>If T_o is any vector space complement to $\zeta_1(L)$ in T then</u>

$$L = \nu(L) \dot{+} T_o$$

<u>and every non-zero element of T_o has a non-zero eigenvalue on L.</u>

Proof: Let $C = C_L(T)$, a Cartan subalgebra of L. By lemma 13.11, $L = \mathcal{V}(L)+C$. Now every ad-nilpotent element of C lies in $\mathcal{V}(L)$ by lemma 14.1, and every ad-pure element lies in T. Hence $L = \mathcal{V}(L) + T$. The rest is clear. □

The next lemma shows that the crucial case to deal with is the locally soluble case:

Lemma 14.7 Let L be ideally finite over k. Then L is cleft if and only if its radical is cleft.

Proof: It is not hard to prove that if L is cleft then so is $\sigma(L)$. For the converse let $S = \sigma(L)$ and suppose S is cleft. Let Λ be a Levi subalgebra of L. Use lemma 14.5 to choose a maximal torus T of S idealized by Λ. By lemma 14.6, $S = N+T$ where $N = \mathcal{V}(S) = \mathcal{V}(L)$. By lemma 14.3(a) there exists a Λ-module complement T_0 to $N \cap T$ in T, and then $S = N \dot{+} T_0$ and T_0 is a torus. Further,

$$[T_0,\Lambda] \subseteq T_0 \cap [S,\Lambda] \subseteq T_0 \cap N = 0$$

by corollary 4.4, so Λ centralizes T_0. Now

$$L = N \dot{+} (T_0 \oplus \Lambda)$$

where T_0 is a torus of L, since it is a torus of S and centralizes Λ. If $x \in L$ then we can write $x = r+t+\ell$ where $r \in N$, $t \in T_0$, $\ell \in \Lambda$. Now Λ is L-cleft by preservation of Chevalley-Jordan decomposition, so we can write $\ell = p+n$ where $p*_L$ is pure, $n*_L$ nilpotent, and $[n,p] = 0$. Now $[t,p] = 0$ so $t+p$ is ad_L-pure.

We claim that $r+n$ is ad_L-nilpotent. It is clear that if k is chosen to make $n^{*^k} = 0$ then $L(r+n)^{*^k} \subseteq N$. Now [64] lemma 2.1 p.15 implies that $N+\langle n\rangle$ is locally nilpotent, and it follows that $r+n$ is ad_N-nil. But $(r+n)^*$ has finite rank, so is nilpotent on N, and there exists j such that $N(r+n)^{*^j} = 0$. Hence $L(r+n)^{*^{k+j}} = 0$ and $r+n$ is ad_L-nilpotent.

Thus we have a semicleaving $x = (r+n)+(t+p)$. We now apply theorem 14.2 to the locally soluble algebra $J = \langle S,n,p\rangle$. We have $J^2 \leq N$. By the remark which follows theorem 14.2, there is an automorphism $\alpha \in \mathcal{E}(J)$ such that

$$x = n' + (t+p)^{\alpha} \qquad (4)$$

is an ad-cleaving on J. Now α extends to an automorphism of L, which we denote by the same symbol $\alpha \in \mathcal{E}(L)$. Since α is a product of terms $\exp(j^*)$ where j lies in a non-zero weight space of J (and hence in J^2, hence in N) we have

$$(t+p)^{\alpha} \equiv t+p \pmod N$$

and hence

$$n' \equiv n \pmod N.$$

The argument that showed $n+r$ ad-nilpotent now shows that n' is ad_L-nilpotent. Since α is an automorphism, $(t+p)^{\alpha}$ is ad_L-pure. Since (4) is an ad-cleaving in J, $[n',(t+p)^{\alpha}] = 0$ and (4) is also an ad-cleaving in L. □

A final preliminary lemma gives a criterion for a split extension to be ideally finite.

Lemma 14.8 Let $L = H \dotplus K$ where $H \triangleleft L$ and H,K are ideally finite over $\mathfrak{k}$. Suppose that H is a locally finite

K-module, and that K acts on H by maps of finite rank. Then L is ideally finite.

Proof: Let I be a K-submodule of H. Then

$$[I,H]K \subseteq [IK,H] + [I,[H,K]] \subseteq [I,H].$$

Hence by induction the ideal closure

$$I^H = I + [I,H] + [I,H,H] + \dots$$

is also a K-module. It follows that every element of H is contained in a finite-dimensional ideal of L.

It remains to prove the same for every element $k \in K$. Now $k \in J \triangleleft K$ where J is finite-dimensional. Since K acts by finite rank maps, $[H,J]$ is finite-dimensional. There is a finite-dimensional K-invariant ideal I of H containing $[H,J]$, and I+J is an ideal of L of finite dimension containing k. □

We come to the fundamental result of this chapter.

Theorem 14.9 Every ideally finite Lie algebra over k can be embedded in a cleft ideally finite Lie algebra.

Proof: Let L be ideally finite, and let $\Gamma(L)$ be the algebra of finite rank derivations of L defined in chapter 13. As in theorem 13.8 we have a homomorphism $*: L \to \Gamma(L)$ whose image L^* is Inn(L) and whose kernel is $Z = \zeta_1(L)$. Let $\Gamma = \Gamma(L)$ and denote images under * by asterisks. Let $S = \sigma(L)$, and choose a Levi subalgebra Λ of L. Then $L^* = S^* \dot{+} \Lambda^*$ and $S^* \cong S/Z$. Now Γ is cleft ideally finite by theorem 13.8. Further $S^* \triangleleft \Gamma$ so $S^* \leq \sigma(\Gamma)$ which is also cleft. By lemma 14.5 there exists a maximal torus T of $\sigma(\Gamma)$ idealized by Λ^* (look at L^*). By the definition of Γ, T acts as a Lie algebra of derivations of S, and we may form the split extension

$$Y = S \dot{+}_{\Gamma} T$$

where the subscript Γ indicates the action of T on S. For clarity we shall denote elements of Y by pairs (s,t) where $s \in S$, $t \in T$.

We claim that Y is cleft. Now if $s \in S$ then $s^* \in \sigma(\Gamma)$. By lemma 14.6 $s^* = r+t$ where $r \in \mathcal{V}(\Gamma)$, $t \in T$. Now r is nil in its action on $S^* = S/Z$, so acts nilpotently on S. The element (s,-t) of Y acts trivially on Y/S, and acts on S as s^*-t which is nilpotent. Hence (s,-t) is ad-nil on Y. For any $u \in T$ we have

$$(s,u) = (s,-t) + (0,u+t) \tag{5}$$

where (0,u+t) acts purely on S and centralizes T, hence acts purely on Y. Hence (5) is a semicleaving, and by theorem 14.2, Y is cleft (it is ideally finite by lemma 14.8).

Since Λ^* idealizes T it follows that $T+\Lambda^*$ is a subalgebra of Γ. The split extension

$$X = S \dot{+}_{\Gamma} (T+\Lambda^*)$$

is ideally finite by lemma 14.8, and its radical is obviously Y which we have seen is cleft. By lemma 14.7, X is cleft. It is easy to verify that the subalgebra $S \dot{+}_{\Gamma} \Lambda^*$ of X is isomorphic to $S \dot{+} \Lambda = L$, so that L embeds in X. □

15 The cleft envelope

Having embedded a given ideally finite Lie algebra L in a cleft ideally finite Lie algebra L', the next step is to try to make L' as small as possible in the hope of obtaining a uniquely determined object. This we shall do in this chapter. We define a cleft envelope of L to be a cleft ideally finite Lie algebra E such that whenever $L \leq E' < E$, E' is not cleft. Mal'cev [46] proves the existence of a cleft envelope (called a 'splitting' in his paper) for any finite-dimensional L over $\mathfrak{k}$. This is an easy consequence of an embedding theorem like that in the previous chapter. In the infinite-dimensional case a little more care is needed to prove that a cleft envelope exists. We begin by proving that *if* it exists, it is unique.

Lemma 15.1 *Let $H_1 \leq K_1$, $H_2 \leq K_2$ be cleft ideally finite Lie algebras over $\mathfrak{k}$, and let $\phi: H_1 \to H_2$ be an isomorphism. Suppose that $x \in H_1$ is ad-cleft in K_1, with $x = x_p + x_n$, where $x_p \in K_1 \setminus H_1$; suppose further that $y = \phi(x)$ is ad-cleft in K_2 with $y = y_p + y_n$, $y_p \notin H_2$. Then there exists an isomorphism $\psi: \langle H_1, x_p, x_n \rangle \to \langle H_2, y_p, y_n \rangle$ extending ϕ.*

Proof: We have $[x_n, x_p] = 0$, $x_n = x - x_p$, and x_p^* is a polynomial in x^* with zero constant term. Hence

$$\langle H_1, x_p, x_n \rangle = H_1 + \langle x_p \rangle$$

and $\langle x_p \rangle$ idealizes H_1. Similarly

$$\langle H_2, y_p, y_n \rangle = H_2 + \langle y_p \rangle$$

and $\langle y_p \rangle$ idealizes H_2.

Let $H_{1,\lambda}$ be the λ-weight space of x^* on H_1. Then for $h \in H_{1,\lambda}$ we must have

$$[h, x_p] = hx_p{}^* = \lambda h.$$

Now $\phi(H_{1,\lambda}) = H_{2,\lambda}$ where the latter is the λ-weight space for $y = \phi(x)$; so it follows that

$$[\phi(h), y_p] = \phi(h)y_p{}^* = \lambda\phi(h).$$

Since the $H_{1,\lambda}$ span H_1 it it clear that if we set $\varphi(x_p) = y_p$ and make $\varphi|_{H_1} = \phi$ then φ is the required isomorphism. □

Using the obvious Zorn's lemma argument this implies:

<u>Theorem 15.2</u> <u>Let H_1 be ideally finite over k, $\phi: H_1 \rightarrow H_2$ an isomorphism, and let K_1 and K_2 be respectively cleft envelopes of H_1, H_2. Then ϕ extends to an isomorphism $\varphi: K_1 \rightarrow K_2$. In particular, a cleft envelope is unique up to isomorphism.</u>

□

The existence of a cleft envelope may be stated in the following strong form:

<u>Theorem 15.3</u> <u>Let L be ideally finite over k, K cleft ideally finite, and let $L \leq K$. Then there exists a cleft envelope E of L with $L \leq E \leq K$.</u>

<u>Proof</u>: Let $I = I_K(L)$. The polynomial property of cleavings implies that I is K-cleft, hence cleft, so without loss of generality $K = I$ and so $L \triangleleft K$. Let $S = \sigma(L)$, $\Sigma = \sigma(K)$. Then $S \triangleleft K$ and $S \leq \Sigma$. Let Λ be a Levi subalgebra of L. Then $\Sigma + \Lambda$ is cleft since Σ is, and $L = S + \Lambda \leq \Sigma + \Lambda$, so without loss of

generality $K = \Sigma+\Lambda$. Let T be a maximal torus of Σ such that Λ idealizes T. Then S+T is a subalgebra of K idealized by Λ, and the argument of theorem 14.9 shows that S+T is cleft, hence $S+T+\Lambda$ is cleft. Without loss of generality $K = S+T+\Lambda$. There is a Λ-invariant complement T_1 to $S \cap T$ in T, so that $S+T = S \dot{+} T_1$. There is also a Λ-invariant complement T_o to $C_{T_1}(S)$. Now $S+T_o \cong (S+T_1)/C_{T_1}(S)$ is cleft, so $S + T_o + \Lambda$ is cleft. Hence without loss of generality

$$K = S \dot{+} T_o \dot{+} \Lambda.$$

Now if $x \in [T_o,\Lambda] \subseteq T_o \cap \mathcal{V}(S+T_o)$ then x is ad-pure and ad-nilpotent, hence x is central, hence $x = 0$ since we have removed $C_{T_1}(S)$. Therefore $[T_o,\Lambda] = 0$. Hence also, every subspace of K lying between S and $S+T_o$ is an ideal.

Now we resort to a process of transfinite induction. We well-order $S+T_o$, and for ordinals α define $S_o = S$; $S_\lambda = \bigcup_{\alpha<\lambda} S_\alpha$ if λ is a limit ordinal; and $S_{\alpha+1} = S_\alpha$ if S_α is cleft, whereas if S_α is not cleft, $S_{\alpha+1} = S_\alpha + \langle x_p \rangle$ where x is the smallest element in the well-ordering such that x^* is not cleft on S_α, $x = x_p+x_n$ being a cleaving in $S+T_o$. The S_α form a well-ordered ascending chain, and on cardinality grounds $S_{\beta+1} = S_\beta$ for some β. Obviously S_β is cleft and contains S: we show it is a cleft envelope for S and that $S_\beta+\Lambda$ is a cleft envelope for L.

We claim that $\mathfrak{z}_1(S_{\alpha+1}) = \mathfrak{z}_1(S_\alpha)$ for all $\alpha < \beta$. For let $z \in \mathfrak{z}_1(S_{\alpha+1})$. If $z \notin S_\alpha$ then $z = s+\lambda x_p$ $(0 \neq \lambda \in k)$. On S_α we have $s^* = -\lambda x_p^*$. Consider the decomposition

$$x = (x+\lambda^{-1}s) + (\lambda^{-1}s).$$

in S_α. Then $-\lambda^{-1}s^* = x_p^*$ is pure, and $(x+\lambda^{-1}s)^* = x_n^*$ is nilpotent. Therefore x is semicleft in S_α, and the remark following theorem 14.2 shows that x is cleft in S_α. Hence $S_{\alpha+1} = S_\alpha$ by definition, contradicting $z \notin S_\alpha$. Therefore $z \in S_\alpha$ and it follows that $z \in \zeta_1(S_\alpha)$.

On the other hand, if $y \in \zeta_1(S_\alpha)$ then $[y,x] = 0$ and the polynomial property implies that $[y,x_p] = 0$ so that y lies in $\zeta_1(S_{\alpha+1})$. This proves the claim.

It follows that $\zeta_1(S_\beta) = \zeta_1(S)$. A similar appeal to the polynomial property of cleavings shows that if $u \in C_K(S_\alpha)$ then $u \in C_K(S_{\alpha+1})$, and in particular $C_{S_\beta}(S) = \zeta_1(S)$.

Now we show that no subalgebra H of S_β with $S \leq H < S_\beta$ is cleft. Let γ be the least ordinal $\leq \beta$ such that $S_\gamma \not\leq H$. Then γ is not a limit ordinal, so $\gamma = \alpha+1$ for some $\alpha < \beta$. We have $S_\alpha \leq H$, $S_{\alpha+1} \leq H$. Let x be the element whose cleaving $x = x_p+x_n$ defines $S_{\alpha+1} = S_\alpha + \langle x_p \rangle$. Then there is an ad-cleaving $x = h_p+h_n$ in H. Now

$$x^* = h_p^* + h_n^* = x_p^* + x_n^*$$

are two cleavings of x^*, thought of as a linear transformation of S_α. Therefore

$$h_p - x_p \in C_{S_\beta}(S_\alpha) = \zeta_1(S)$$

so that $h_p \in S_{\alpha+1}$, and $S_{\alpha+1} \leq H$. This is a contradiction. Therefore S_β is a cleft envelope for S.

Finally let $J \leq S_\beta+\Lambda$ where $L = S+\Lambda \leq J \leq S_\beta+\Lambda$. If J is cleft then $\sigma(J)$ is cleft and $S \leq \sigma(J) \leq S_\beta$, so $\sigma(J) = S_\beta$. Hence $J = S_\beta+\Lambda$ and the latter is a cleft envelope for L. □

In conjunction with theorem 14.9 and lemma 15.1 this result implies:

Theorem 15.4 Every ideally finite Lie algebra L has a cleft envelope which is unique up to isomorphism (extending the identity map on L). Any monomorphism from L to a cleft ideally finite Lie algebra extends to a monomorphism of the cleft envelope. □

We shall denote the cleft envelope of L by $\tilde{L}$, and identify L with its image as a subalgebra of $\tilde{L}$. We now show that the structures of L and $\tilde{L}$ are intimately related, a result which allows us to use $\tilde{L}$ as a tool to elucidate the structure of L.

Theorem 15.5 Let L be ideally finite over k with cleft envelope $\tilde{L}$. Then:

(a) Every ideal of L is an ideal of $\tilde{L}$.

(b) $\tilde{L}^2 = L^2$.

(c) $\zeta_1(\tilde{L}) = \zeta_1(L)$.

(d) $\tilde{L} = \nu(\tilde{L}) + L$.

(e) $\nu(\tilde{L}) = \nu(L)$ if and only if $\tilde{L} = L$.

(f) $\sigma(\tilde{L})$ is a cleft envelope for $\sigma(L)$.

(g) $\tilde{L} = \sigma(\tilde{L}) + \Lambda$ where Λ is any Levi subalgebra of L.

(h) Every automorphism of L extends, uniquely modulo $\zeta_1(L)$, to an automorphism of $\tilde{L}$.

Proof: (a) The subalgebra $\bigcap_{I \triangleleft L} I_{\tilde{L}}(I)$ is cleft, by the polynomial property, hence must be all of $\tilde{L}$.

(b) Define $\tilde{L}$ in terms of the S_α as in theorem 15.3. By transfinite induction $S_\alpha^{\ 2} = S^2$ and $[S_\alpha,\Lambda] = [S,\Lambda]$ for all α, and with $\tilde{L} = S_\beta+\Lambda$ we get $\tilde{L}^2 = L^2$.

(c) This has already been established for S_β.

(d) We claim that $\nu(\tilde{L})+L$ is cleft. For if $r \in \nu(\tilde{L})$, $x \in L$ then $x = r'+t$ where $r' \in \nu(L)$ and t is ad-pure; so that $r+x = (r + r') + t$ is a semicleaving of x in $\nu(\tilde{L})+ L$. Minimality implies $\tilde{L} = \nu(\tilde{L})+L$.

(e) This follows from (d).

(f) and (g) are true by the proof of theorem 15.3.

(h) The existence is clear by theorem 15.4. Uniqueness modulo the centre follows from uniqueness, modulo the centre, of ad-cleavings. □

In [46] Mal'cev asserts uniqueness of the extended automorphism in (h), but this is not true, as is shown by an example of Mostow, Chevalley, and Jacobson (cf. [46] p.261).

16 Maximal locally nilpotent subalgebras

As one application of the cleft envelope we shall derive an analogue, for ideally finite Lie algebras, of Mal'cev's classification of maximal nilpotent subalgebras. Of course the results are less strong, and we do not get his finiteness theorem ([46] p.231). First we sharpen lemma 14.6.

Lemma 16.1 *Let L be cleft locally soluble ideally finite over k, and let H be an L-cleft subalgebra. If U is any torus of L, maximal with respect to $U \leq H$, then*

$$H = (H \cap \nu(L)) + U$$

and

$$(H \cap \nu(L)) \cap U = H \cap \zeta_1(L).$$

If U_0 is any vector space complement to $H \cap \zeta_1(L)$ in U then

$$H = (H \cap \nu(L)) \dot{+} U_0$$

and the non-zero elements of U_0 have non-zero eigenvalues on L.

Proof: If $h \in H$ then $h = h_0 + h_1 + \ldots + h_r$ where

$$[h_i, u] = \lambda_i(u) h_i \qquad (1)$$

for distinct eigenvalues $\lambda_i : U \to k$, where $\lambda_0 = 0$. If $i > 0$ then (1) implies that $h_i \in H \cap L^2 \leq H \cap \nu(L)$. Since $\lambda_0 = 0$ we have $h_0 \in C_H(U)$. Hence

$$H = (H \cap \nu(L)) + C_H(U).$$

Suppose that

$$H \neq H_1 = (H \cap \nu(L)) + U.$$

Choose $c \in C_H(U) \setminus H_1$, and let $c = c_p + c_n$ be an ad-cleaving in L, such that $c_p, c_n \in H$. Then $c_n \in \nu(L) \cap H$. If $c_p \in H_1$ then $c \in H_1$, which is a contradiction, so $c_p \notin H_1$. But now $U + \langle c_p \rangle$ is a torus of L, contained in H, contradicting the maximality of U. Hence $H = H_1$. All else is obvious. □

Theorem 16.2 Let L be cleft locally soluble ideally finite over k, with $L = N \dot{+} T_0$, $N = \nu(L)$, T_0 a torus. Then up to $\mathcal{L}(L)$-conjugacy every maximal locally nilpotent subalgebra M of L has the form

$$M = C_N(T_1) \dot{+} T_1$$

where $T_1 \leq T_0$ has the property that $C_{T_0}(C_N(T_1)) = T_1$.

Proof: It is not hard to show that M must be L-cleft. By lemma 16.1, $M = (N \cap M) \dot{+} U_0$ for a torus U_0 of L. Let $T = T_0 + Z$, where $Z = \zeta_1(L)$. Then T is a maximal torus, so there exists $\alpha \in \mathcal{L}(L)$ such that $U_0^{\alpha} \leq T$, and

$$M^{\alpha} = (N \cap M^{\alpha}) \dot{+} U_0^{\alpha}.$$

Now $M \geq Z$, so we may choose U_0 to make $U_0^{\alpha} \leq T_0$. Put $T_1 = U_0^{\alpha}$. Then $M^{\alpha} = (N \cap M^{\alpha}) \dot{+} T_1$ and $T_1 \leq T_0$. Now T_1 is a torus, and since M is locally nilpotent, the eigenvalues of elements of T_1 on M^{α} are all zero, and $[M^{\alpha}, T_1] = 0$. Thus the (obviously locally nilpotent) subalgebras

$$C_N(T_1)+T_1 , \qquad C_T(N \cap M^{\alpha})+(N \cap M^{\alpha})$$

contain M^{α}. Maximality implies firstly that $M^{\alpha} = C_N(T_1)+T_1$, secondly that $C_T(C_N(T_1)) = T_1$, as claimed.

It is easy to verify that algebras of the type described are maximal locally nilpotent. □

The next theorem allows us to reduce to the cleft case from the locally soluble case.

Theorem 16.3 Let L be locally soluble ideally finite over k, with cleft envelope $\tilde{L}$. Then every maximal locally nilpotent subalgebra of L is contained in a unique maximal locally nilpotent subalgebra of $\tilde{L}$. The $\mathcal{L}(L)$-conjugacy classes of maximal locally nilpotent subalgebras of L are in bijection with a set of $\mathcal{L}(\tilde{L})$-conjugacy classes of maximal locally nilpotent subalgebras of $\tilde{L}$.

Proof: The strongly nilpotent elements of L lie in L^2. By theorem 15.5(b) $\tilde{L}^2 = L^2$, so the strongly nilpotent elements of L and $\tilde{L}$ coincide. We may therefore identify the groups $\mathcal{L}(L)$ and $\mathcal{L}(\tilde{L})$.

Let $\tilde{L} = N \dot{+} T_o$ as usual, and let H be a maximal locally nilpotent subalgebra of L. There is a unique minimal $\tilde{L}$-cleft subalgebra H_1 of $\tilde{L}$ containing H (by an argument similar to the proof of theorem 15.3, noting that $\zeta_1(\tilde{L}) = \zeta_1(L) \leq H$). A transfinite induction shows that H_1 is locally nilpotent. Hence it lies inside a maximal locally nilpotent subalgebra H_2 of $\tilde{L}$. By theorem 16.2 we have

$$H_2 = C_N(T_1) \dot{+} T_1$$

where, for some $\alpha \in \mathcal{L}(\tilde{L})$, $T_1^{\alpha} \leq T_o$.

Let $t \in T_1$. We can find $r \in N$ such that $t+r \in L$, by theorem 15.5(d). Put $r = r_o + r_1 + \dots + r_s$ where

$$[r_i, u] = \lambda_i(u) r_i$$

for all $u \in T_1$, where the $\lambda_i : T_1 \to k$ are distinct eigenvalues, and $\lambda_0 = 0$. Now $r_1, \ldots, r_s \in \tilde{L}^2 = L^2$, so $r_0 + t \in L$. Hence $r_0 \; C_N(T_1)$, and $r_0 + t \in H_2 \cap L = H$ by maximality. Therefore $t \in H_1$ by definition of H_1, and $T_1 \leq H_1$.

We claim that there is a unique torus U of $\tilde{L}$ maximal with respect to $U \leq H_1$. For if U, U' have this property then they are central in H_1 by local nilpotency, so $[U,U'] = 0$ and U+U' is a torus.

Now it is easy to see that $T_1 + \zeta_1(L)$ is a torus of L, maximal subject to being contained in H_1. The previous paragraph shows T_1 to be unique modulo $\zeta_1(L)$, so H_2 is uniquely determined.

Since $H_2 \cap L = H$ by maximality, we have an injection from the set of maximal locally nilpotent subalgebras of L to that of $\tilde{L}$. The identification of $\mathcal{L}(L)$ with $\mathcal{L}(\tilde{L})$ leads to an induced injection on the corresponding conjugacy classes. $\square$

By Mal'cev [46] theorem 5 p.238 it follows that if t is ad-pure, L is locally soluble, $\gamma \in \mathcal{L}(L)$, and $[t, t^{\gamma}] = 0$, then $t = t^{\gamma}$. Hence for a fixed maximal torus T of L, every torus is $\mathcal{L}(L)$-conjugate to a unique subtorus of T. Hence the description in theorem 8.2 yields inconjugate subalgebras provided that it yields unequal ones. To obtain a complete description in this case, all we need is a characterization of those subtori T_1 of T_0 with the property $C_{T_0}(C_N(T_1)) = T_1$. It is easy to prove that these are precisely the subtori of T_0 of the from $C_{T_0}(R)$, for $R \leq N$. In fact there is a 'Galois correspondence' between the lattice of subalgebras of N and the lattice of subalgebras of T_0, with maps

$$R \mapsto C_{T_0}(R) \qquad (R \leq N)$$

$$T \mapsto C_N(T) \qquad (T \leq T_0)$$

which restrict to a bijection between tori of the form $C_{T_0}(R)$ and subalgebras of the form $C_N(T)$.

This opens the way to a more useful description in terms of Mal'cev's concept of 'type'. Let L be cleft, locally soluble, ideally finite over $\mathfrak{k}$. By theorem 3.6 we can find an ascending series of ideals $(L_\alpha)_{\alpha \leq \sigma}$ with factors $L_{\alpha+1}/L_\alpha$ of dimension 1. If we decompose L as usual, $L = \mathcal{V}(L) \dot{+} T_0$, then there is a basis of T_0-eigenvectors adapted to the series.

Fix such a series. Each $x \in L$ acts on $L_{\alpha+1}/L_\alpha$ by scalar multiplication $\lambda_\alpha(x)$. We define an ordering of the elements of L by setting $x \gg y$ if for all $\alpha, \beta \leq \sigma$,

$$\lambda_\alpha(x) \neq \lambda_\beta(x) \text{ implies } \lambda_\alpha(y) \neq \lambda_\beta(y).$$

The relation $\sim$ defined by

$$x \sim y \iff x \gg y \text{ and } y \gg x$$

is an equivalence relation, and the <u>type</u> of x is defined to be its equivalence class. Then $\gg$ induces a partial order on the types, which we denote by the same symbol.

By looking at a basis of T_0-eigenvectors adapted to the series $(L_\alpha)_{\alpha \leq \sigma}$ it is easy to see that for $x \in T_0$ the centralizer $C_L(x)$ depends only on the type of x. Hence for $R \leq N$ the subtorus $T_1 = C_{T_0}(R)$ is a union of types, and contains along with each type all smaller types. Hence we obtain:

Theorem 16.4 Let L be cleft locally soluble ideally finite over k. With the above notation, two subtori T_1 and T_2 of T_o define $\mathcal{L}(L)$-conjugate maximal locally nilpotent subalgebras of L if and only if neither of T_1, T_2 has an element of type strictly greater than the type of every element of the other. □

We can pass from the locally soluble case to the general case in the following manner: fix a Borel subalgebra B of L. Then every maximal locally nilpotent subalgebra M of L has some $\mathcal{L}(L)$-conjugate M^{α} inside B. Now $\mathcal{L}(B)$-conjugacy implies $\mathcal{L}(L)$-conjugacy, so there is a surjection from the set of all $\mathcal{L}(B)$-conjugacy classes of maximal locally nilpotent subalgebras of B, classified as above, to the set of $\mathcal{L}(L)$-conjugacy classes of maximal locally nilpotent subalgebras of L.

There remains the question: is this map injective? We have not decided this. However, the following proposition may be relevant:

Proposition 16. 5 Let L be cleft ideally finite over . If B is a Borel subalgebra of L there is a decomposition

$$B = \mathcal{V}(B) + V$$

where V is a maximal torus of L, every element of $\mathcal{V}(B)$ is ad-nilpotent on L, and $\mathcal{V}(B) \cap V = \zeta_1(L)$.

Proof: Let $S = \sigma(L)$ and choose a Levi factor Λ. If D is a Borel subalgebra of Λ then S+D is a Borel subalgebra of L, which must be $\mathcal{L}(L)$-conjugate to B. Changing Λ if necessary we may assume B = S+D. We can decompose D as $R \dotplus U_o$ where

U_o is a torus of Λ, $R = \nu(D)$, and the non-zero elements of U_o act on Λ with non-zero eigenvalues (lemma 16.1). The theory of finite-dimensional semisimple Lie algebras shows that the elements of $\nu(D)$ are ad_Λ-nilpotent. Preservation of Chevalley-Jordan decomposition shows that U_o is a torus of L and the elements of $\nu(D)$ are ad_L-nilpotent.

Decompose S into $N \dot{+} T_o$ where $N = \nu(S)$ and T_o is a Λ-invariant torus, such that non-zero elements of T_o act on S with non-zero eigenvalues. Then we have a decomposition

$$B = (N \dot{+} R) \dot{+} (T_o \oplus U_o)$$

because U_o must centralize T_o, by the usual argument. Then T_o centralizes Λ and so is a torus of L, and it follows that $T_o \oplus U_o$ is a torus of L. Also, $N \dot{+} R$ is locally nilpotent since elements of R are ad_L-nilpotent, and it is not hard to check that elements of $N \dot{+} R$ are ad-nilpotent on L.

To see that $N \dot{+} R$ is $\nu(B)$, note the assertions about non-zero eigenvalues of T_o and U_o.

Let V be a torus of L, maximal subject to $U_o \oplus T_o \leq V \leq B$. Then $B = \nu(B) + V$ and clearly $\nu(B) \cap V = \zeta_1(L)$. It remains to show that V is a maximal torus of L. Suppose $V \leq W$, where W is a torus of L. Then an argument like that above allows us to quotient by S and hence assume L semisimple, and it is then immediate from the structure theory that $W = V$. □

17 Intravariance

The concept of an intravariant subalgebra of a Lie algebra was introduced by Barnes [5] by analogy with finite group theory, to allow a substitute for the 'Frattini argument'. In the ideally finite case a modified concept is more appropriate. Let $\mathfrak{X}$ be any class of Lie algebras. Let H be a subalgebra of a Lie algebra L. Suppose that whenever there exists an $\mathfrak{X}$-algebra X with $L \triangleleft X$ we have

$$X = L + I_X(H).$$

Then we say that H is <u>$\mathfrak{X}$-intravariant</u> (or <u>intravariant relative to $\mathfrak{X}$</u>) in L. When $\mathfrak{X}$ is the class of all Lie algebras this reduces to intravariance. We want the case when $\mathfrak{X}$ is the class of ideally finite Lie algebras: reference to $\mathfrak{X}$ will be tacit, using the phrase <u>relatively intravariant</u>.

<u>Lemma 17.1</u> <u>Let L be ideally finite over k. Then $H \leq L$ is relatively intravariant if and only if every derivation of L of finite rank is the sum of an inner derivation and a derivation stabilizing H.</u>

<u>Proof</u>: Assume the condition on derivations fulfilled. Let $H \triangleleft X$ where X is ideally finite. If $x \in X$ then x^* leaves L invariant, and $x^*|_L$ is a derivation of finite rank. Hence

$$x^*|_L = y^*|_L + \delta$$

where $y \in L$ and $H\delta \subseteq H$. Then $x-y \in I_X(H)$ and $x \in L+I_X(H)$.

Conversely, suppose H relatively intravariant in L, and let δ be a derivation of finite rank. Then L is a locally finite $\langle\delta\rangle$-module, so by lemma 14.8 the split extension $L \dot{+} \langle\delta\rangle$

is ideally finite. Hence $\delta = x+y$ where $x \in L$ and y idealizes H. In terms of derivations, $\delta = x^*|_L + y^*|_L$ and $y^*|_L$ stabilizes H. $\square$

Barnes [5] shows that with suitable hypotheses, if $\mathcal{H}$ is a homomorph of Lie algebras and H is an $\mathcal{H}$-projector, then H is intravariant. Analogously, taking for $\mathcal{H}$ the classes of locally soluble, locally nilpotent, and semisimple Lie algebras, we have:

<u>Theorem 17.2</u> <u>Let L be ideally finite over k. Then the Borel, Cartan, and Levi subalgebras of L are all relatively intravariant.</u>

<u>Proof</u>: For Cartan subalgebras this is lemma 14.4. Let B be a Borel subalgebra of $L \triangleleft X$ where X is ideally finite. Let $S = \sigma(L) \triangleleft X$. Then $S \leq B$. We may quotient out S, or equivalently assume L semisimple. Then L is a direct summand of X by proposition 6.4, so $X = L + C_X(L) = L + C_X(B)$. In the original situation, $X = L + I_X(B)$ as required.

For Levi subalgebras we prove a formally stronger result: that every semisimple subalgebra is relatively intravariant. However, since any characteristic ideal of an intravariant subalgebra is itself intravariant, the apparent generality is largely spurious.

Let S be a semisimple subalgebra of L, and let δ be a derivation of finite rank. We use lemma 17.1. Suppose first that S has finite dimension. There exists a finite-dimensional δ-invariant subalgebra F of L such that $S \leq F$, and $\delta|_F$ is a derivation of S into F. By Jacobson [39] p.80 theorem 9 there exists $f \in F$ such that $f^*|_S = \delta|_S$. Then $f^*-\delta$ annihilates S, and $\delta = f^* + (\delta-f^*)$ shows that S is relatively intravariant.

Finally we remove the dimension hypothesis. Now we have $S = S_o \oplus S_1$ where S_o has finite dimension and $S_1\delta = 0$. Then $S_o \leq C = C_L(S_1)$ and C is δ-invariant. By the finite-dimensional case there exists $c \in C$ such that $c^*|_{S_o} = \delta|_{S_o}$. But $c^*|_{S_1} = \delta|_{S_1} = 0$, so $c^*|_S = \delta|_S$. Hence $\delta = c^*+(\delta-c^*)$ where c^* is inner and $\delta-c^*$ stabilizes S. □

18 Cartan subalgebras of ideals

Winter [78] has proved a number of theorems relating Cartan subalgebras of ideals to Cartan subalgebras of the whole Lie algebra (pp. 115-128, especially theorem 4.4.5.3, p.128). We shall extend some of his results to ideally finite Lie algebras in this chapter, omitting routine extensions of theorems proved in the literature in finite dimensions.

If M is a locally finite N-module, where N is locally nilpotent, we shall write $M_0(N)$ for the 0-weight space of M. Let L be ideally finite over k. We say that a subalgebra N of Der(L) is <u>quasiregular</u> if

(i) N is locally nilpotent,

(ii) L is a locally finite N-module,

(iii) $L_0(N)$ is locally nilpotent.

A subalgebra of L is quasiregular if the induced algebra of derivations is: in this case (ii) is automatically verified.

<u>Proposition 18.1</u> <u>Let</u> L <u>be ideally finite over</u> k. <u>A subalgebra</u> N <u>of</u> L <u>is quasiregular if and only if</u> $L_0(N)$ <u>is a Cartan subalgebra of</u> L.

<u>Proof:</u> Winter [78] 4.4.2.6 p.118. □

<u>Theorem 18.2</u> <u>Let</u> L <u>be ideally finite over</u> k. <u>For a locally nilpotent subalgebra</u> C <u>of</u> L <u>the following are equivalent:</u>

(a) C <u>is a Cartan subalgebra</u>,

(b) $C = L_0(C)$,

(c) C <u>is a maximal quasiregular subalgebra of</u> L.

Proof: By theorem 13.10, (a) and (b) are equivalent. To show that (b) $\Rightarrow$ (c) $\Rightarrow$ (a) argue as in Winter [78] 4.4.2.9 p.119. □

A Fitting subalgebra of L is a subalgebra of the form $L_o(N)$ where N is a locally nilpotent subalgebra of Der(L) for which L is a locally finite module. If further N is the image under the adjoint map of a locally nilpotent subalgebra of L then we say $L_o(N)$ is an Engel subalgebra.

Theorem 18.3 Let L be ideally finite over k. If E is an Engel subalgebra of L and C is a Cartan subalgebra of E, then C is a Cartan subalgebra of L.

Proof: We use corollary 10.5. Let K be any ideal of L of finite codimension. Then E+K/K is an Engel subalgebra of L/K and C+K/K is a Cartan subalgebra of E+K/K. Hence by the finite-dimensional version of the theorem in Winter [78] 4.4.4.8 p.124, C+K/K is a Cartan subalgebra of L/K. Clearly C contains the centre of L, since E does, and it follows from 10.5 that C is a Cartan subalgebra of L. □

Corollary 18.4 Every Engel subalgebra of L contains a Cartan subalgebra of L, and the Cartan subalgebras of L are precisely the minimal Engel subalgebras. □

(This is a generalization of Winter [78] 4.4.4.7 p.124. A direct proof along the same lines does not seem to work, since we have no analogue of his theorem 4.3.2 p.109).

Proposition 18.5 Let L be ideally finite over k. Then every quasiregular subalgebra N of L is contained in a unique Cartan subalgebra

$$N^{\dagger} = L_0(N)$$

of L.

Proof: Winter [78] 4.4.2.8 p.118. □

Proposition 18.6 Let L be ideally finite over k. Then any Cartan subalgebra C of a Fitting subalgebra of L is quasi-regular, hence contained in a unique Cartan subalgebra $C^{\dagger}$ of L.

Proof: The only difficulty is in showing that $L_0(C)$ is locally nilpotent: this comes from corollary 10.5 and Winter [78] 4.4.4.10 p.125. □

If L is ideally finite over k, and I is an ideal, we can define for a Cartan subalgebra C of L,

$$C|_I = I_0(C \cap I).$$

Lemma 18.7 With the above notation, $C|_I$ is a Cartan subalgebra of I.

Proof: Let $N = I_0(C) = L_0(C) \cap I = C \cap I$. Then N is quasi-regular. Now $C|_I = I_0(N) = N^{\dagger}$ (in I) which is a Cartan subalgebra of I by 18.5. □

Probably the main result of this section is:

Theorem 18.8 Let L be ideally finite over $\mathfrak{k}$, having an ideal I. The map $C \to C|_I$ is a surjection from the set of Cartan subalgebras C of L to the set of Cartan subalgebras of I. For a given Cartan subalgebra D of I, the inverse image of D under this map is the set of Cartan subalgebras of $L_o(D)$.

Proof: $\mathcal{L}(I)$-conjugacy of Cartan subalgebras of I, and the extension property of $\mathcal{L}$, shows the map surjective. The rest follows as in Winter [78] 4.4.5.3 p.128, using the next lemma. □

Lemma 18.9 For a Cartan subalgebra C of L and a Cartan subalgebra D of I, the following are equivalent:

(a) $D = C|_I$,

(b) C idealizes D,

(c) $C \leq L_o(D)$,

(d) $C \cap I \leq D$.

Proof: Winter [78] 4.4.5.2 p.127. □

19 Locally soluble algebras

In this section we generalize some theorems of Stitzinger [66] to locally soluble ideally finite Lie algebras, and apply the results to clarify the connection between Cartan subalgebras of an ideally finite Lie algebra and its cleft envelope. We shall not consider analogues of Stitzinger's 'system normalizers' since in characteristic zero these coincide with the Cartan subalgebras.

<u>Proposition 19.1</u> *Let L be locally soluble ideally finite over $\mathfrak{k}$, and let H be a subalgebra of L such that $L = \nu(L)+H$. Each Cartan subalgebra K of H is contained in a unique Cartan subalgebra C of L. Further $C \cap H = K$ and $C \geq I_L(K)$.*

<u>Proof:</u> We adapt Stitzinger [66] theorem 4. Let $S = L_0(K)$ which contains K since K is locally nilpotent. If we can show that S is locally nilpotent then K is quasiregular, so by proposition 18.1 S is a Cartan subalgebra. Let $N = \nu(L)$. Now $K+N/N$ is a Cartan subalgebra of $H+N/N = L/N$, so is equal to its 0-weight space $S+N/N$ by theorem 18.2. Hence $K+N = S+N$, so $S = S \cap (K+N) = K+(S \cap N)$ which is locally nilpotent since $S \cap N \lhd S$ and K acts nilpotently on S. The uniqueness of Cartan subalgebras containing K follows from proposition 18.5. Now $S \cap H$ is locally nilpotent and contains K, so equals K by maximality. Obviously S contains $I_L(K)$. □

We can apply this to the relation between L and $\tilde{L}$ when L is locally soluble (as a preliminary to the general case).

Corollary 19.2 Let L be locally soluble ideally finite over $\mathfrak{k}$. Then every Cartan subalgebra K of L is uniquely expressible as $C \cap L$ where C is a Cartan subalgebra of $\tilde{L}$. The map $C \mapsto C \cap L$ is a bijection between the sets of Cartan subalgebras of $\tilde{L}$ and of L.

Proof: We know that $\tilde{L} = \nu(\tilde{L})+L$ by theorem 15.5(d), so by the above every Cartan subalgebra of L is uniquely of the form $C \cap L$ for a Cartan subalgebra of $\tilde{L}$. As in theorem 16.3 we may identify $\mathcal{L}(L)$ and $\mathcal{L}(\tilde{L})$, and it follows that the given map is bijective. □

A quick 'bootstrap' extends this to the general case:

Theorem 19.3 Let L be ideally finite over $\mathfrak{k}$. Let $\tilde{L}$ be its cleft envelope, T a maximal torus of $\tilde{L}$. Then $C_L(T)$ is a Cartan subalgebra of L and all Cartan subalgebras of L arise in this way. The map $C \mapsto C \cap L$ is a bijection between the sets of Cartan subalgebras of $\tilde{L}$ and of L, compatible with $\mathcal{L}(\tilde{L})$- or $\mathcal{L}(L)$-conjugacy.

Proof: By theorem 13.2, $C = C_{\tilde{L}}(T)$ is a Cartan subalgebra of $\tilde{L}$. Let B be a Borel subalgebra of $\tilde{L}$ containing C. By corollary 8.3 there is a Levi subalgebra Λ of $\tilde{L}$ and a Borel subalgebra B_Λ of Λ such that $B = \sigma(\tilde{L})+B_\Lambda$. By 15.5(f) we have $\sigma(\tilde{L}) = \widetilde{\sigma(L)}$. If $B_1 = \sigma(L)+B_\Lambda$ then B_1 is a Borel subalgebra of L, and $\tilde{B}_1 = B$. By corollary 19.2, $C \cap B_1$ is a Cartan subalgebra of B_1, hence of L by lemma 11.2. But

$$C \cap B_1 = C_{B_1}(T) \leq C_L(T) \leq C_L(T)$$

which is locally nilpotent. Hence $C_{B_1}(T) = C_L(T)$ is a Cartan subalgebra of L. □

Other results of Stitzinger carry over to ideally finite Lie algebras. We give brief indications of the proofs.

Proposition 19.4 *Let L be locally soluble ideally finite over* $\mathfrak{k}$. *Then the following are equivalent for a subalgebra* C *of* L:

(a) C *is a Cartan subalgebra of* L,

(b) C *is a maximal locally nilpotent subalgebra of* L *and* $L = \nu(L) + C$,

(c) C+K/K *is a maximal locally nilpotent subalgebra of* L/K *for every ideal* K *of* K,

(d) C *contains the centre of* L *and* C+K/K *is a maximal locally nilpotent subalgebra of* L/K *for all ideals* K *of* L *such that* dim L/K *is finite*.

Proof: It is clear that (a) implies each of (b), (c), (d), and that (c) implies (d). Assume (b): then C is contained in a unique Cartan subalgebra of L by 19.1, and this must be C. To show (c) implies (d) reduce to finite dimensions by corollary 10.5 and use theorem 5 of Stitzinger [66]. □

Proposition 19.5 *Let L be ideally finite over* $\mathfrak{k}$, *having a series of ideals* $N_i \triangleleft L$,

$$0 = N_0 \leq N_1 \leq \cdots \leq N_n = L,$$

such that the factors N_{i+1}/N_i are all locally nilpotent. Then C is a Cartan subalgebra of L if and only if $C+N_i/N_i$ is maximal locally nilpotent in L/N_i for $i = 0,\ldots,n-1$.

Proof: Argue as in Stitzinger [66] theorem 5(2), but using lemma 10.1(d) instead of Barnes [3] lemma 4. □

Now we consider 'cover-avoidance' properties. Let L be ideally finite over $\mathfrak{k}$, with $H \leq L$. If M,N are H-submodules of L and $N \triangleleft M$ then M/N is an H-factor of L. If M/N is irreducible as H-module then it is an irreducible H-factor. It is central if $[M,H] \leq N$, eccentric otherwise. We say that H avoids M/N if $N+(M \cap H) = N$, and covers M/N if $N+(M \cap H) = M$.

Proposition 19.6 Let L be ideally finite locally soluble over $\mathfrak{k}$. Then C is a Cartan subalgebra of L if and only if C covers every central, and avoids every eccentric, C-factor of L.

Proof: Lemma 4 and theorem 6 of Stitzinger [66] show that a Cartan subalgebra has the stated cover-avoidance property. For the converse, refer to the finite-dimensional quotients of L using corollary 10.5 and apply theorem 7 of Stitzinger [66]. □

20 Complementation theorems

It is easy to see that every locally finite Lie algebra L has a unique ideal I minimal with respect to L/I being locally nilpotent. This is the locally nilpotent residual $\lambda(L)$.

We introduce some notation. If N is a locally nilpotent Lie algebra, M a locally finite N-module, we write $M_{\alpha}(N)$ for the weight space with weight α, and put

$$M_*(N) = \bigoplus_{\alpha \neq 0} M_{\alpha}(N).$$

If C is a Cartan subalgebra of L then $L = \lambda(L)+C$ by the projector property. Sometimes we can assert more: C is a complement to $\lambda(L)$. For instance, we have an analogue of theorem 5.15 of Carter and Hawkes [13].

Theorem 20.1 Let L be ideally finite over $\mathfrak{k}$ and suppose that $\lambda(L)$ is abelian. Then $\lambda(L)$ is complemented in L and the complements are precisely the Cartan subalgebras. The complements are $\mathcal{L}(L)$-conjugate.

Proof: Let C be a Cartan subalgebra of L, and $R = \lambda(L)$. Then $C \cap R = R_0(C)$ is complemented by $R_*(C)$. Now C idealizes $R_*(C)$ by chapter 12, formula (4), and R idealizes it (being abelian) so $R_*(C) \lhd L$. Now $L = R_*(C) \dotplus C$ and $L/R_*(C)$ is locally nilpotent, so $R_*(C) = R$. Hence R is complemented by any Cartan subalgebra.

Next suppose K complements R. Then K is locally nilpotent and R is abelian so $R \leq \nu(L)$. By proposition 19.1 there is a Cartan subalgebra $C \geq K$. But C complements R as well, so $C = K$. Conjugacy is obvious. □

Pursuing this idea, we say that an ideally finite Lie algebra L is Cartan-complemented if there exists an ideal I and a Cartan subalgebra C such that $L = I \dotplus C$. By theorem 20.1 every abelian-by-locally nilpotent Lie algebra L (i.e. having an abelian ideal with locally nilpotent quotient) is Cartan-complemented. Borel subalgebras of semisimple algebras are Cartan-complemented (using (b) of the next theorem) but are not in general abelian-by-locally nilpotent.

Theorem 20.2 *For an ideally finite Lie algebra L over k the following are equivalent:*

(a) *L is Cartan-complemented,*

(b) *$L_*(C)$ is a subalgebra of L for some (and hence every) Cartan subalgebra C of L,*

(c) *For some (and hence every) Cartan subalgebra C of L and every weight $\alpha : C \to k$,* $[L_\alpha(C), L_{-\alpha}(C)] = 0$.

Proof: We know that $L_*(C)$ is the unique C-module complement to $C = L_0(C)$ in L. The rest follows easily using formula (4) of section 12. □

Theorem 20.3 *Let L be a Cartan-complemented ideally finite Lie algebra over k, with $L = I \dotplus C$ for an ideal I and a Cartan subalgebra C. Then*

(a) *$L = I \dotplus D$ for any Cartan subalgebra D.*

(b) *$I = \lambda(L) = L_*(D)$ for any Cartan subalgebra D.*

(c) *I is locally nilpotent and L is locally soluble.*

(d) *The complements to I are precisely the Cartan subalgebras of L.*

<u>Proof</u>: (a) is immediate by $\mathcal{L}(L)$-conjugacy. For (b), there is a unique D-module complement $L_*(D)$ to $D = L_0(D)$ so $L_*(D) = I$. Now $\lambda(L) \leq I$ since L/I is locally nilpotent, but $L = \lambda(L)+C$, hence $\lambda(L) = I$. For (c), theorem 18.8 implies that $I_0(I \cap C) = I_0(0) = I$ is a Cartan subalgebra of I, hence I is locally nilpotent. Part (d) follows as in theorem 20.1. □

Theorem 20.1 can be extended as in Stitzinger [66] theorem 8, replacing $\lambda(L)$ by $\lambda^n(L)$ defined recursively by $\lambda^{i+1}(L) = \lambda(\lambda^i(L))$, $\lambda^1(L) = \lambda(L)$. Stitzinger's notation is L^n_∞. There is an $\mathcal{L}(L)$-conjugacy result for the complements.

Appendix: Fitting classes

A dualization of formation theory, the theory of Fitting classes of finite soluble groups, was introduced by Fischer [17]. We mentioned in the Introduction to these notes that the corresponding dualization to soluble finite-dimensional Lie algebras seems less fruitful. The purpose of this appendix is to support this contention.

Throughout this appendix, all Lie algebras considered will be finite-dimensional and soluble, over a field of characteristic zero. Algebraic closure of the field will not be required.

A Fitting class of (soluble finite-dimensional) Lie algebras is a class $\mathcal{B}$ such that

If $I, J \triangleleft L$ and $I, J \in \mathcal{B}$ then $I+J \in \mathcal{B}$, (1)

If $I \triangleleft L \in \mathcal{B}$ then $I \in \mathcal{B}$. (2)

It is easy to verify (cf. [54]) that if $\mathcal{B}$ is a Fitting class every Lie algebra L has a unique maximal $\mathcal{B}$-ideal, which we call the $\mathcal{B}$-radical and denote by $\rho_{\mathcal{B}}(L)$. This is a characteristic ideal of L, and contains every $\mathcal{B}$-subideal of L.

We say that a Lie algebra L is atomic, or is an atom, if L is not the sum of two proper ideals. The Fitting class generated by a set of Lie algebras is the smallest Fitting class containing that set.

Proposition 1 Every Fitting class is generated by its atoms.

Proof: Let $\mathcal{B}$ be a Fitting class, $\mathcal{X}$ the class of all atoms $L \in \mathcal{B}$, and let $\mathcal{D}$ be the Fitting class generated by $\mathcal{X}$. Obviously $\mathcal{D} \subseteq \mathcal{B}$. For the reverse inclusion we prove that if $L \in \mathcal{B}$ then $L \in \mathcal{D}$, by induction on dim L. Let dim L = n. If $n \leq 1$ then L is an atom, and the result holds. If n > 1, either L is an atom and the result holds, or L = I+J where I,J are proper ideals. By induction I and J lie in $\mathcal{D}$, and since $\mathcal{D}$ is a Fitting class $L \in \mathcal{D}$ as required. □

Corollary 2 Every non-zero Fitting class contains the class $\mathfrak{N}$ of all nilpotent algebras.

Proof: If $\mathcal{B}$ is a non-zero Fitting class then $\mathcal{B}$ contains a Lie algebra $L \neq 0$, and since L is soluble it has a subideal of dimension 1. Hence $\mathcal{B}$ contains the Fitting class generated by its 1-dimensional subalgebras.

If L is a nilpotent atom then it is easy to verify that dim $L \leq 1$. For if dim L > 1 we can find a proper ideal I of codimension 1. If $x \notin I$ then it generates an ideal x^L which is not the whole of L (by nilpotency), and then L is the sum of I and x^L so is not atomic.

Proposition 2 now implies that $\mathfrak{N} \subseteq \mathcal{B}$. □

Proposition 3 A Lie algebra L is atomic if and only if either dim $L \leq 1$ or L is a split extension $N \dot{+} \langle d \rangle$ where N is a nilpotent ideal and d induces a nonsingular derivation on N/N^2.

Proof: If dim $L \leq 1$ the result is obvious. Suppose that

L is atomic and dim L $\geq$ 2. Since any decomposition of L/L^2 as a sum of two proper subspaces lifts to a decomposition of L as a sum of two proper ideals, it follows that dim $L/L^2 = 1$. Let $L^2 = N$, and pick $d \in L \setminus N$. Then $L = N \dot{+} \langle d \rangle$. Since $N = L^2$, N is nilpotent.

Since L is atomic the ideal d^L generated by d must be the whole of L, so every element of N is a sum of products of terms of the form

$$[d, n_1, \ldots, n_r]$$

where $n_1, \ldots, n_r \in N$. It follows that d acts on N/N^2 as a nonsingular transformation.

Conversely, suppose that $L = N \dot{+} \langle d \rangle$ where d acts as a nonsingular derivation of N/N^2, and N is a nilpotent ideal. Suppose for a contradiction that L is a sum of two proper ideals I and J. If $L \neq N+I$ then $L = (N+I)+J$ is another such decomposition, so we may assume $N \leq I$; if $L = N+I$ this decomposition allows us to assume $N \leq I$ as well. Now J must contain an element of the form d+n where $n \in N$. Now d acts nonsingularly on N/N^2 while n annihilates it; hence d+n acts nonsingularly on N/N^2, and it follows that

$$\langle (d+n)^N \rangle + N^2 = L.$$

Now it is easy to see ([1] lemma 2.1.9 p.40) that this implies $(d+n)^N = L$, so that $J = L$, a contradiction. □

If Inn(L) is the algebra of inner derivations of L and Out(L) = Der(L)/Inn(L), then Aut(L) can be made to act upon Out(L) as follows. For $d \in \mathrm{Der}(L)$, $\alpha \in \mathrm{Aut}(L)$ define

$$d^{\alpha} = \alpha^{-1} d \alpha \in \mathrm{Der}\,(L).$$

Since Inn(L) is invariant under this action, Aut(L) acts on Out(L). The action commutes with scalar multiplication, so Aut(L) acts on the associated projective space Pout(L). We call an orbit of Aut(L) in Pout(L) nonsingular if it arises from a derivation of L which is nonsingular on L/L^2. When L is nilpotent inner derivations are nilpotent, so that any derivation lying in a nonsingular orbit is nonsingular.

Proposition 4 *Let* N *be a nilpotent Lie algebra. Then there is a bijection between:*

(a) *Isomorphism classes of atomic Lie algebras* L *such that* $L^2 = N$,

(b) *Nonsingular orbits of* Aut(N) *in its action on* Pout(N).

Proof: Given two nonsingular derivations d and e of N, we must find the condition for isomorphism of $N \dot{+} \langle d \rangle$ and $N \dot{+} \langle e \rangle$. If α is such an isomorphism then $\alpha|_N$ is an automorphism of N, and there exists $n \in N$, $0 \neq \lambda$ (a scalar), such that $d\alpha = \lambda e + n$. It follows that for all $x \in N$,

$$[x,d]\alpha = [x\alpha, \lambda e+n] = [x\alpha, \lambda e] + [x\alpha, n]$$

so that

$$[x,d] = [x, \lambda e\alpha^{-1}] + [x, n\alpha^{-1}]$$

and d and e lie in the same Aut(N) orbit on Der(L), modulo scalar multiples and inner automorphisms. Their images in Pout(L) thus lie in the same orbit.

The converse is clear. □

In the special case where N is abelian then Aut(N) is GL(N) and Out(N) = gl(N), and the orbit problem reduces to classical problems of linear algebra.

The analogue, for a Fitting class $\mathcal{B}$, of the projectors of formation theory is defined as follows. If L is a Lie algebra then a $\mathcal{B}$-<u>injector</u> of L is a subalgebra V such that, for any subideal S of L, $V \cap S$ is a maximal $\mathcal{B}$-subalgebra of S. This is not a useful concept, at least in characteristic zero, because of:

<u>Proposition 5</u> <u>For each Fitting class</u> $\mathcal{B}$ <u>every Lie algebra L has a unique</u> $\mathcal{B}$<u>-injector, which is equal to the</u> $\mathcal{B}$<u>-radical of</u> L.

<u>Proof</u>: We may assume $\mathcal{B} \neq (0)$. Let N be the nil radical of L. By corollary 2, $N \in \mathcal{B}$. If V is any $\mathcal{B}$-injector of L, then $V \cap N$ is a maximal $\mathcal{B}$-subalgebra of N, so $V \cap N = N$, and $N \leq V$. Therefore $V \triangleleft L$ since $N \geq L^2$, and therefore $V \leq \rho_{\mathcal{B}}(L)$. But $V \cap \rho_{\mathcal{B}}(L)$ must equal $\rho_{\mathcal{B}}(L)$, so $V = \rho_{\mathcal{B}}(L)$.

Conversely, since $\rho_{\mathcal{B}}(L) \cap S = \rho_{\mathcal{B}}(S)$ for each subideal S of L (cf. [54]), it follows that $\rho_{\mathcal{B}}(L)$ is a $\mathcal{B}$-injector. □

In group theory, the injectors form a single conjugacy class. Here so much more is true that the theory becomes largely trivial.

However, there is some interest in exhibiting Fitting classes, since these seem to be somewhat rare. Let $\mathfrak{N}$, $\mathcal{S}$ be the classes of nilpotent, soluble algebras (respectively). We define three more classes as follows. $L \in \mathfrak{M}$

if every minimal ideal is central. $L \in \mathfrak{T}$ if every inner derivation of L has trace zero on L. $L \in \mathfrak{U}$ if every inner derivation has trace zero when restricted to any minimal ideal of L.

Theorem 6 The classes $0, \mathfrak{N}, \mathfrak{S}, \mathfrak{M}, \mathfrak{T}, \mathfrak{U}, \mathfrak{M} \cap \mathfrak{T}$ are all Fitting classes, and are distinct. Inclusions between them are given by the lattice

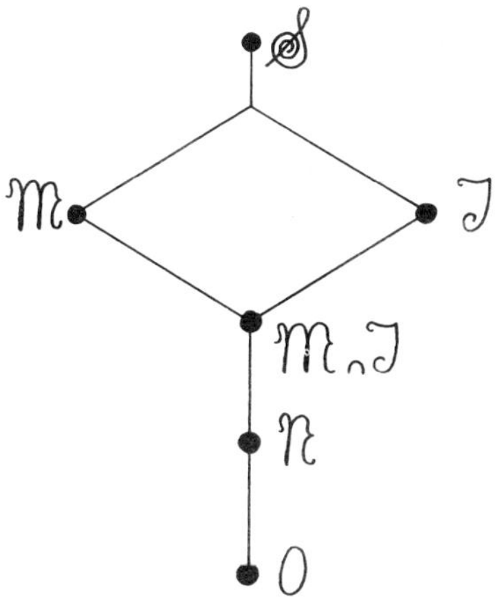

Proof: All inclusions are easy. It is well known that $0, \mathfrak{N}, \mathfrak{S}$ are Fitting classes. Since intersections of Fitting classes are obviously Fitting classes, it remains to prove that $\mathfrak{M}, \mathfrak{T}, \mathfrak{U}$ are Fitting classes.

The class $\mathfrak{T}$: Let $L = I+J$ where $I, J \triangleleft L$ and $I, J \in \mathfrak{T}$. Clearly L acts by trace zero maps on $I \cap J$. Modulo $I \cap J$ we have $[I,J] = 0$, so that each of I and J act by trace zero maps on $L/I \cap J$. Hence L acts by trace zero maps on $L/I \cap J$, and it is easy to see that $L \in \mathfrak{T}$.

Now let $I \triangleleft L \in \mathfrak{T}$. Then I centralizes L/I and acts by trace zero maps on L. Hence it acts by trace zero maps on I, so $I \in \mathfrak{T}$.

The class $\mathfrak{M}$: Let $L = I+J$ where I and J are ideals of L, lying in $\mathfrak{M}$. Let K be a minimal ideal of L. Then $K \cap I = 0$ or $K \leq I$, so I centralizes K. Similarly J centralizes K, so K is central in $I+J = L$.

Now let $I \triangleleft L \in \mathfrak{M}$. Let K be a minimal ideal of I. Define $\overline{K} = \Sigma_{\alpha \in \mathrm{Aut}(I)} K^{\alpha}$. By theorem 1.3 $\overline{K}$ is a characteristic ideal of I, so $\overline{K} \triangleleft L$. Hence $\overline{K}$ intersects the centre of L nontrivially. Now $\overline{K}$ is a sum of minimal ideals, hence a direct sum of certain K^{α}. Let x be a non-zero element of $\overline{K}$ in the centre of L. Decomposing with respect to the direct decomposition of $\overline{K}$ we have

$$x = x_1 + \ldots + x_r$$

with each x_r central in I. Some $x_i \neq 0$, so there exists some $y \in K$, $y \neq 0$ such that y^{α} is central in I for an automorphism α of I. Thus y is central in I, and by minimality, K is central.

The class $\mathfrak{U}$: The proof is similar to the above, using obvious properties of the trace.

This establishes that the classes described are Fitting classes. We show the seven classes described are distinct as follows.

$\mathfrak{U} \neq \mathfrak{S}$: the algebra $\langle a,b \mid [a,b] = a \rangle$ does not lie in $\mathfrak{U}$.

$\mathfrak{T} \cap \mathfrak{M} \neq \mathfrak{M}$: Let M be the Lie algebra of 4x4 matrices, and let P be the subalgebra of zero-triangular matrices. Let d be a diagonal matrix with diagonal terms $(0,-1,0,0)$. Then d acts as a derivation on P. Let $L = \langle P,d \rangle = P \dot{+} \langle d \rangle$. Any minimal ideal of L must intersect the centre of the nil radical P, and this has dimension 1 and is annihilated by d. Therefore $L \in \mathfrak{M}$. However, the trace of d* on L is 1, so $L \notin \mathfrak{T}$.

$\mathfrak{T} \neq \mathfrak{T} \cap \mathfrak{M}$: Let

$$L = \langle x, y, z, d \mid [xy] = z,\ [xd] = x,\ [yd] = -y,\ [zd] = 0 \rangle .$$

Then $L \in \mathfrak{T}$ but $L \notin \mathfrak{M}$.

All other inequalities in the diagram follows from these. □

For $\mathfrak{T}$ and $\mathfrak{U}$, the radical has codimension ≤ 1. It is perhaps surprising that distinct Fitting classes can have radicals so close to the whole algebra.

The seven examples above can all be viewed as variations on a single theme: the action of L on its chief factors. Thus

$L \in \mathfrak{N}$ iff all chief factors are central

$L \in \mathfrak{T}$ iff all chief factors are 'trace-zero'

$L \in \mathfrak{M}$ iff all minimal ideals are central

$L \in \mathfrak{U}$ iff all minimal ideals are 'trace-zero'.

Whether this observation is indicative of a general phenomenon we must leave unanswered here.

References

1. R.K.Amayo and I.N.Stewart: Infinite-dimensional Lie algebras, Noordhoff International, Leyden 1974.

2. R.Baer: Sylow theorems for infinite groups, Duke Math. J. 6 (1940) 598-614.

3. D.W.Barnes: On Cartan subalgebras of Lie algebras, Math. Z. 101 (1967) 350-355.

4. - : Lie algebras (lecture notes), University of Tübingen 1968-69.

5. - : The Frattini argument for Lie algebras, Math. Z. 133 (1973) 277-283.

6. D.W.Barnes and H.M. Gastineau-Hills: On the theory of soluble Lie algebras, Math. Z. 106 (1968) 343-354.

7. D.W.Barnes and M.L.Newell: Some theorems on saturated homomorphs of soluble Lie algebras, Math. Z. 115 (1970) 179-187.

8. S.G.Brazier: Stability and parasolubility of Lie rings, Ph.D. thesis, University of Warwick 1973.

9. A.Borel: Linear algebraic groups, Benjamin, New York 1969.

10. N.Bourbaki: Topologie générale I (3rd edition), Hermann, Paris 1961.

11. - : General topology part 1, Addison-Wesley, Reading Massachusetts 1966.

12. R.W.Carter: Nilpotent self-normalizing subgroups of soluble groups, Math. Z. 75 (1961) 136-139.

13. R.W.Carter and T.O.Hawkes: The $\mathfrak{F}$-normalizers of a finite soluble group, J. Algebra 5 (1967) 175-202.

14. C.Chevalley: Theorie des groupes de Lie, tome II: groupes algébriques, Hermann, Paris 1951.

15. C.W.Curtis: On Lie algebras of algebraic linear transformations, Pacific J. Math. 6 (1956) 453-466.

16. M.Demazure and P.Gabriel: Groupes algébriques, tome I, North-Holland, Amsterdam 1970.

17. B.Fischer: Habilitationsschrift, University of Frankfurt.

18. B.Fischer, W.Gaschütz, and B.Hartley: Injectoren endlicher auflösbarer Gruppen, Math. Z. 102 (1967) 337-339.

19. W.Gaschütz: Zur Theorie der endlichen auflösbaren Gruppen, Math. Z. 80 (1963) 300-305.

20. P.A.Gol'berg: Sylow π-subgroups of locally normal groups, Mat. Sb. 19 (1946) 451-460 (Russian).

21. C.J.Graddon and B.Hartley: Basis normalizers and Carter subgroups in a class of locally finite groups, Proc. Cambridge Philos. Soc. 71 (1972) 189-198.

22. P.Hall: A note on soluble groups, J. London Math. Soc. (1) 3 (1928) 99.

23. - : On the Sylow systems of a soluble group, Proc. London Math. Soc. (2) 43 (1937) 316-323.

24. - : On the system normalizers of a soluble group, Proc. London Math. Soc. (2) 43 (1937) 507-528.

25. P.Hall: Theorems like Sylow's, Proc. London Math. Soc. (3) 6 (1956) 286-304.

26. - : Periodic FC-groups, J. London Math. Soc. (1) 34 (1959) 289-304.

27. B.Hartley: Locally nilpotent ideals of a Lie algebra, Proc. Cambridge Philos. Soc. 63 (1967) 257-272.

28. - : $\mathfrak{F}$-abnormal subgroups of certain locally finite groups, Proc. London Math. Soc. (3) 23 (1971) 128-158.

29. - : Sylow subgroups of locally finite groups, Proc. London Math. Soc. (3) 23 (1971) 159-192.

30. - : Some examples of locally finite groups, Arch. der Math. 23 (1972) 225-231.

31. - : Sylow theory in locally finite groups, Compositio Math. 25 (1972) 263-280.

32. - : Sylow p-subgroups and local p-solubility, J. Algebra 23 (1972) 347-369.

33. - : A note on $\mathfrak{F}$-reducibility, J. London Math. Soc. (2) 6 (1972) 161-168.

34. - : A class of modules over a locally finite group I, J. Austral. Math. Soc. 26 (1973) 431-442.

35. - : A class of modules over a locally finite group II, (to appear).

36. - : On Fischer's dualization of formation theory, Proc. London Math. Soc. (3) 19 (1969) 193-207.

37. B.Hartley, A.D.Gardiner, and M.J.Tomkinson: Saturated formations and Sylow structure in locally finite

groups, J. Algebra 17 (1971) 177-211.

38. J.E.Humphreys: Introduction to Lie algebras and representation theory, Graduate Texts in Mathematics 9, Springer, Berlin 1972.

39. N.Jacobson: Lie algebras, Interscience, New York 1962.

40. I.Kaplansky: Lie algebras an locally compact groups, Chicago 1971.

41. O.H.Kegel and B.A.F.Wehrfritz: Locally finite groups, North Holland, 1972.

42. A.A.Klimowicz: Fitting and formation theory in locally finite groups, Ph.D. thesis, University of Warwick 1973.

43. - : Formation theory in locally finite groups, Proc. London Math. Soc. (3) 30 (1975) 257-286.

44. - : Sylow structure and basis normalizers in a class of locally finite groups, (to appear).

45. A.G.Kuroš: Theory of groups, vol. 2, Chelsea, New York 1956.

46. A.I.Mal'cev: Solvable Lie algebras, Izv. Akad. Nauk SSSR 9 (1945) 329-352 (Russian); Amer. Math. Soc. Translations Ser. 1 vol. 9, Lie groups, 1962, 229-262.

47. E.I.Marshall: The Frattini subalgebra of a Lie algebra, J. London Math. Soc. 42 (1967) 416-422.

48. D.H.McLain: A characteristically simple group, Proc. Cambridge Philos. Soc. 50 (1954) 641-642.

49. M.L.Newell: Homomorphs and formats in polycyclic groups, J. London Math. Soc. (2) 7 (1973) 317-327.

50. - : The nilpotent-by-finite projectors of polycyclic groups, J. London Math. Soc. (2) 7 (1973) 540-546.

51. H.Samelson: Notes on Lie algebras, Van Nostrand Reinhold Mathematical studies 23, New York 1969.

52. E.Schenkman: A theory of subinvariant Lie algebras, Amer. J. Math. 73 (1951) 433-474.

53. J.-P. Serre: Groupes proalgébriques, Publ. Math. I.H.E.S. 7 (1960) 341-403.

54. I.N.Stewart: Structure theorems for a class of locally finite Lie algebras, Proc. London Math. Soc. (3) 24 (1972) 79-100.

55. - : Levi factors of infinite-dimensional Lie algebras, J. London Math. Soc. (2) 5 (1972) 488.

56. - : Author-abstract of [54], Zbl. 225 (1972) 109.

57. - : Conjugacy theorems for a class of locally finite Lie algebras, Compositio Math. (to appear).

58. - : Chevalley-Jordan decomposition for a class of locally finite Lie algebras, (to appear).

59. - : The structure of certain infinite-dimensional Lie algebras, Proc. 3rd International Colloquium on group-theoretical methods in physics, Marseille 1974, 474-483.

60. I.N.Stewart: The Wedderburn-Mal'cev theorems in a locally finite setting, Arch. der Math. (to appear).

61. - : The Lie algebra of endomorphisms of an infinite-dimensional vector space, Compositio Math. 25 (1972) 79-86.

62. - : The minimal condition for subideals of Lie algebras, Math. Z. 111 (1969) 301-310.

63. - : Infinite-dimensional Lie algebras in the spirit of infinite group theory, Compositio Math. 22 (1970) 313-331.

64. - : Lie algebras, Lecture Notes in Mathematics 127, Springer, Berlin 1970.

65. I.N.Stewart and D.A.Towers: The Frattini subalgebras of certain infinite-dimensional soluble Lie algebras, J. London Math. Soc. (to appear).

66. E.L.Stitzinger: Theorems on Cartan subalgebras like some on Carter subgroups, Trans. Amer. Math. Soc. 159 (1971) 307-315.

67. - : Covering-avoidance for saturated formations of solvable Lie algebras, Math. Z. 124 (1972) 237-249.

68. S.E.Stonehewer: Abnormal subgroups of a class of periodic locally soluble groups, Proc. London Math. Soc. (3) 14 (1964) 520-536.

69. - : Locally soluble FC-groups, Arch. der Math. 16 (1965) 158-177.

70. S.E.Stonehewer: Formations and a class of locally soluble groups, Proc. Cambridge Philos. Soc. 62 (1966) 613-635.

71. M.J.Tomkinson: Formations of locally soluble FC-groups, Proc. London Math. Soc. (3) 19 (1969) 675-708.

72. - : $\mathfrak{F}$-injectors of locally soluble FC-groups, Glasgow Math. J. 10 (1969) 130-136.

73. D.A.Towers: A Frattini theory for algebras, Proc. London Math. Soc. (3) 27 (1973) 440-462.

74. W.Tuck: A Frattini theory for Lie algebras, Ph.D. thesis, University of Sydney 1969.

75. F.-H.Vasilesçu: On Lie's theorem in operator algebras, Trans. Amer. Math. Soc. 172 (1972) 365-372.

76. B.A.F.Wehrfritz: Sylow theorems for periodic linear groups, Proc. London Math. Soc. (3) 18 (1968) 125-140.

77. - : Soluble periodic linear groups, Proc. London Math. Soc. (3) 18 (1968) 141-157.

78. D.J.Winter: Abstract Lie algebras, M.I.T. Press, Cambridge Massachusetts 1972.

Index